The City Project

Strategies for Smart and Wise Sustainable Urban Design

Volume 4

The book series The City Project reports on applied research and operational developments that promote urban renewal and the sustainable transformation of contemporary cities. Inspired by the "City of Man" as imagined by Adriano Olivetti and Ernesto Nathan Rogers, and going beyond the concept of the smart city and related technological advances, the series' goal is to present holistic, practice-oriented and multidisciplinary strategies for realizing the City 4.0, i.e., the city of the fourth industrial revolution, in keeping with the objectives of the UN's 2030 Agenda for Sustainable Development.

In particular, the series reports on effective design, planning and management approaches that leverage urban and architectural design skills, engineering, environmental and social expertise, and administrative abilities alike. It welcomes books on each of the aspects mentioned above, as well as studies analyzing multiple aspects, their interactions and/or holistic solutions. The City Project addresses a very broad readership, including designers, engineers, architects, social scientists, stakeholders and public administrators, who deal with various aspects of the realization of the City 4.0. It publishes theoretical investigations into the contemporary built environment, international case studies, and pilot projects concerning urban renewal and the regeneration of urban areas, as well as the proceedings of key international conferences.

Books published in this series are devoted to supporting education, professional training and public administration. Outstanding PhD theses on emerging topics, if properly reworked, may also be considered for publication. The series is published with the support of the Smart City 4.0 Sustainable LAB, an interdisciplinary teaching and research project on future cities initiated by the University of Parma, and jointly implemented with other regional universities (the University of Bologna, University of Ferrara, and University of Modena and Reggio Emilia).

About the Cover

The cover of the book series The City Project features a painting by Carlo Mattioli (C. Mattioli, *Estate in Versilia*, 1974, oil on canvas cm. 118 × 70, Catalog n. 1974D0029, Courtesy of Fondazione Carlo Mattioli, thanks to Anna Zaniboni Mattioli)

The horizon of poppies painted by Carlo Mattioli between the dark background of the forest and the white plane of the wheat, becomes for us, thanks to a transfiguration of meaning that aligns with the attitude towards abstraction rooted in the figure of the painter, a city which is intertwined with its landscape, evoking the idea and the possibility of recomposing a balance and seeking an integration between settlement and environment, between human space and natural element.

Andrea Borsari · Annalisa Trentin ·
Pierpaolo Ascari
Editors

TEMPORARY: Citizenship, Architecture and City

Springer

Editors
Andrea Borsari
Department of Architecture
University of Bologna
Bologna, Italy

Annalisa Trentin
Department of Architecture
University of Bologna
Bologna, Italy

Pierpaolo Ascari
Department of Architecture
University of Bologna
Bologna, Italy

ISSN 2730-6992 ISSN 2730-700X (electronic)
The City Project
Strategies for Smart and Wise Sustainable Urban Design
ISBN 978-3-031-36666-6 ISBN 978-3-031-36667-3 (eBook)
https://doi.org/10.1007/978-3-031-36667-3

The cover of the book series The City Project features a painting by Carlo Mattioli (C. Mattioli, *Estate in Versilia*, 1974, oil on canvas cm. 118 × 70, Catalog n. 1974D0029, Courtesy of Fondazione Carlo Mattioli, thanks to Anna Zaniboni Mattioli)

This Springer imprint is published by the registered company Springer Nature Switzerland AG
The registered company address is: Gewerbestrasse 11, 6330 Cham, Switzerland

Acknowledgments

The volume was published with the contribution of the Department of Architecture of the Alma Mater Studiorum—University of Bologna, Department of Excellence MIUR (L. 232 of 01/12/2016).

ALMA MATER STUDIORUM
UNIVERSITÀ DI BOLOGNA
DIPARTIMENTO DI ARCHITETTURA
DIPARTIMENTO DI ECCELLENZA MIUR
(L.232 DEL 1/12/2016)

Contributions to this volume have been subjected to a double peer-reviewed process.

Contents

On Temporary: Citizenship, Architecture and City 1
Pierpaolo Ascari, Andrea Borsari, and Annalisa Trentin

Exploring Temporary

The Crisis of Citizenship. A Symptom of Societal Destructuration 13
Gregor Fitzi

The Dangerous Adventure of Designing Bubbles 27
Lidia Gasperoni

Whose Shadow? On Camps and Counter-Camps 39
Federico Rahola

Inhabiting the Ecological Conversion: Experiments in Diavolution 49
Andrea Mubi Brighenti

Images of the People. Populus, Plethos, Plebs and Ethnos 57
Jacopo Galimberti

Making the Temporary Become Chronic: The Transit City 65
Pierpaolo Ascari

Navigating Crises. Transient Communities for Urban Preparedness 71
Martina Massari

Energy Community and Citizenship as Enabling Actions for Integrated Energy Plan Implementation and Urban Energy Transition 81
Serena Pagliula

Art of Temporary Living: Looking Inside Student Rooms 91
Arshia Eghbali

Permanent versus Temporary: A Struggle within City Transformations 101
Annalisa Trentin

Dissonant Times 109
Andrea Borsari

For a Lemmary of Temporary Citizenship

Assemblages/Decentralised 123
Andrea Cattabriga

Transition/Deviance 127
Valentina De Matteo

Hacktivism/Sexual Tourism 131
Francesco Di Maio

Life Cycle Thinking/Life Cycle Phase 137
Lorna Dragonetti

Time-Based Packaged Goods/*Valigia* (Suitcase, Luggage) 141
Clara Giardina

Adaptation/*Cueva* and *Tomas De Terreno* 145
Marco Iannantuono

Temporary Uses/Pop-Up Spaces 149
Giulia Marzani

Classroom/Digital Citizenship/Material Culture 153
Angelo Massafra

***Grenze* Versus *Schranke*/Open Form** 159
Claudia Nigrelli

Deliberation/Climate Justice 167
Marco Palma

Border/Rules 171
Dafni Retzepi

Appendix 175

Author Index 177

Editors and Contributors

About the Editors

Andrea Borsari teaches Aesthetics at the Department of Architecture of the University of Bologna, directs the series "La vita e le forme" (Bup) and is deputy director of "Iride. Rivista di filosofia e discussione pubblica" (Il Mulino). He was recently visiting professor at UCLA, ENSA-Paris La Villette, the four cities Unica Euromaster in Urban Studies - University of Copenhagen and the University of Sichuan and visiting fellow at University of Amsterdam, University of Buenos Aires and Brown University - Providence. Among his publications, *Mondo, cose, immagini. Sulle forme dell'esperienza estetica* (Bup, 2018), *Mimicry. Estetica del divenire animale* (Mimesis, 2018) and, as editor, *Philosophical Anthropology and Critical Theory* (Iride, 02/2022), *Aesthetics of the Anthropocene* (CPCL, 1–2/2022, with P. Ascari).

Annalisa Trentin is an architect and full professor in the field of architectural and urban design at the Department of Architecture of the Alma Mater Studiorum—University of Bologna; since 2018 she has been Coordinator of the PhD course in Architecture and Design Cultures (Unibo); from 2015 to 2021, she held the position of Deputy Director of the Department of Architecture, where she has been carrying out research and interdisciplinary teaching activities since 2001. She carries out design research by participating in international workshops and competitions; she is the author of several essays on pedagogical practice in architectural design, criticism and theory of contemporary architecture.

Pierpaolo Ascari teaches Aesthetics at the Department of Architecture, University of Bologna. He is author of *Ebola e le forme* (Rome, 2017), *Attraverso i confini. Lettura, storia ed esperienza estetica in Stendhal e Flaubert* (Milan, 2018), *Corpi e recinti. Estetica ed economia politica del decoro* (Verona, 2019) and *The Adventure of Form: Aesthetics, Nature and Society* (Delft, 2021). He edited the volumes *Oggetti contesi. Le cose nella migrazione* (Milan, 2020), *Espressioni urbane. Muri sconciati, writing e street art* (Milan, 2022, whit Pietro Rivasi) and *I luoghi delle merci* (Milan, 2022).

Contributors

Pierpaolo Ascari Department of Architecture, University of Bologna, Bologna, Italy

Andrea Borsari Department of Architecture, University of Bologna, Bologna, Italy

Andrea Cattabriga Department of Architecture, University of Bologna, Bologna, Italy

Valentina De Matteo Department of Architecture, Advanced Design Unit, University of Bologna, Bologna, Italy

Francesco Di Maio Department of Architecture, University of Bologna, Bologna, Italy

Lorna Dragonetti Department of Architecture, University of Bologna, Bologna, Italy

Arshia Eghbali Department of Architecture, University of Bologna, Bologna, Italy;
Department of Arts and Cultural Studies, University of Copenhagen, Copenhagen, Denmark

Gregor Fitzi Faculty for Social and Cultural Sciences, European University Viadrina, Frankfurt (Oder), Germany

Jacopo Galimberti Department of Cultura del Progetto, IUAV, University of Venice, Venice, Italy

Lidia Gasperoni Department of Architectural Theory, Technical University Berlin, Berlin, Germany

Clara Giardina Department of Architecture, University of Bologna, Bologna, Italy

Marco Iannantuono Department of Architecture, University of Bologna, Bologna, Italy

Giulia Marzani Department of Architecture, University of Bologna, Bologna, Italy

Angelo Massafra Department of Architecture, University of Bologna, Bologna, Italy

Martina Massari Department of Architecture, University of Bologna, Bologna, Italy

Andrea Mubi Brighenti Department of Sociology, University of Trento, Trento, Italy

Claudia Nigrelli Department of Architecture, University of Bologna, Bologna, Italy

Serena Pagliula Department of Architecture, University of Bologna, Bologna, Italy

Marco Palma Department of Architecture, University of Bologna, Bologna, Italy

Federico Rahola Dipartimento Di Scienze Della Formazione, Università Di Genova, Genova, Italy

Dafni Retzepi Department of Architecture, University of Bologna, Bologna, Italy

Annalisa Trentin Department of Architecture, University of Bologna, Bologna, Italy

On Temporary: Citizenship, Architecture and City

Pierpaolo Ascari, Andrea Borsari, and Annalisa Trentin

1. This book is the result of a research track that for about two years involved doctoral students, researchers and professors from the Department of Architecture at the University of Bologna and its Doctoral School on Architecture and Design Cultures. The goal of the research group was to try to understand and define the spurious forms of citizenship represented first and foremost by the city life of migrants, off-site students and tourists. In the following text, the various articulations of the work on temporary forms of citizenship that has been carried out are first described (§2), then the possible specifications of the theme of temporarity in relation to the urban-architectural dimension, to the temporary of the city (§3), and to the very forms of temporality with which it is related (§4) are shown. Finally, the contents of the volume are briefly presented (§5).[1]
2. The preliminary exploration of the field thus revealed the presence of two perspectives of analysis among which it could prove fruitful to include our contribution. The first was that relating to so-called partial citizenship, that is, the way in which exceptions to the limits of formal citizenship are introduced among different levels of governance to enable those without it to still access a range of services provided directly by municipal governments. To this area (which has been addressed, for example, by the BABELS research program, directed by Michel Agier at the EHESS in Paris) the studies on ville-refuge in France or Sanctuary cities in the United States and Canada belong [1]. A second attempt at conceptualization, on the other hand, was provided by research on performative citizenship, which, starting from J. L. Austin's theory of linguistic acts, proposed to shift attention to the effects of citizenship that are determined in the context of mobilizations, disputes and conflicts [2].

What seemed to us to emerge from both perspectives was the historical and theoretical

P. Ascari (✉) · A. Borsari (✉) · A. Trentin (✉)
Department of Architecture, University of Bologna, Bologna, Italy
e-mail: pierpaolo.ascari@unibo.it

A. Borsari
e-mail: a.borsari@unibo.it

A. Trentin
e-mail: annalisa.trentin@unibo.it

[1] The text of this introduction is the result of the joint work of the editors of the volume, the writing of the individual paragraphs should be attributed: for §1 and §5 to all three, for §2 to Pierpaolo Ascari, for §3 to Annalisa Trentin and for §4 to Andrea Borsari.

A. Borsari et al. (eds.), *TEMPORARY: Citizenship, Architecture and City*, The City Project 4,
https://doi.org/10.1007/978-3-031-36667-3_1

problem of a material citizenship that even in the absence of recognition is made effective by its own expression on the level of uses and production of space. On the basis of these premises, therefore, we set up a program of in-depth study that would eventually lead to the construction of a lemmary by the doctoral students, the holding of a workshop in which the doctoral students would be able to discuss their papers, and the holding of an international conference. In the meantime, the presentation of the individual research tracks (civic uses, architectural experiments, ways of living with risk, shared apartments and decarbonization processes) allowed us to identify a promising point of convergence on the dimension of "practices," depending on the hypothesis that the status of citizenship within the changes (in relation to migration flows, touristification, the pandemic emergency and the environmental issue) should not immediately refer to a reconceptualization of the term, but rather to the analysis of the concrete forms in which everyday life and institutions can adhere to the transformations that are affecting the relationships between the built and lived city. From practices, in other words, we felt that it was possible to extrinsicate what are the directions in which the status of citizenship is being reconfigured.

Uses and practices inform change, in fact, just as the forms in which change is determined can only foster new concrete experiences of space. As far as we were concerned, then, the aesthetic-perceptual sphere could operate in this dialectic in the dual guise of production and product, as if to say that something is always happening in the city that concerns the precedence of bodies and their ability to tacitly direct what happens on the frontier of urban transformation.

In particular, our research has posed the problem of these orientations in relation to their manifestations in perceived space through the notion of mark, to which at least four meanings are attributed in the literature: the meaning of perceptual mark and operational mark that enable the definition of an environment from the perspective of biology and animal studies [3]; the meaning that the mark takes on in the production and related deconstruction of gender and race hierarchies [4]; the meaning that the mark is attributed to at the origin of tourist studies [5]; and the mark understood as the representation of a power relationship [6, 7]. In the latter meaning—which perhaps encompasses them all—the expression of forms of temporary citizenship through the marking of spaces (or the production of spaces through their use) can only be determined in correspondence with the so-called "contact zones" between the different actors contending for the urban scene [8], assigning to the same centrality of conflict postulated by the theorists of performative citizenship a more immediately physical and sensory significance. To the definition of temporary citizenship, therefore, we felt we could associate the analysis of the way in which individuals and networks shape the space they experience by converting it to their own needs and living conditions.

3. City and architecture are dominated by a long-term planning process, but increasingly the city embraces strategies to include new needs dictated by societal transformations. The city becomes an open system where the dimension of the temporality, especially at a time when scarcity of resources and uncertainty, drive design towards medium-term solutions: temporary solutions for underused buildings, abandoned or 'waiting' areas, become an important tool to develop initiatives and to drive change towards new uses in a sustainable way.

The city, by its nature, is a constantly changing organism and is characterized by the coexistence of contrasting elements: from monuments of the consolidated city to shopping centers, from industrial areas to service areas. In planning processes, temporary solutions are very often only considered to testify a moment of crisis or an error in the development and where transience is rarely seen as an explicit and vital characteristic of cities or as a key component for growth and regeneration [9]. In the urban context, transitional solutions are mainly related to the uncertainty of the scenario, the uncertainty of the model to be used and the uncertainty of the decision-making process; some of these sources of uncertainty are related to differences in the world view, others to the limits of our current knowledge, and some others to our ability to represent processes within urban development models, including the reliability of model input data on spatial and temporal scales [10]. Now we can see a large numbers of experiments, with the temporary use of informal urban spaces, that are ranging from small infill sites to former industrial sites, from former housing developments, often on demolished housing sites, to abandoned infrastructure, including areas between railway lines and abandoned roads, and including empty and unrenovated shops, offices and flats, parks and gardens, including green spaces without a clear identity. The uncertain situation of these spaces, mentioned above as an example, leads, through their temporary use, towards solutions related to the world of art and culture with new uses, including art installations, with the reuse of spaces for studios, workshops and performance spaces, towards sport and recreation, towards entrepreneurship, with the development of spaces for start-ups, open-air markets, bars and towards alternative dwelling situations.

It's in playing with that relationship between the formal and the informal city that there is a window of opportunity for changing people's attitudes towards the unknown and the uncertainty [11].

Finally, temporary use defines a new approach to urban development. Temporary becomes a way that can activate new possibilities of use for forgotten space rather than regulating it, through generic legislation that cannot be applied to specific cases. This suggests us that it is necessary to look more at what is in place instead of approaching the urban structure as a tabula rasa, taking more seriously a design made through small steps and through the contributions of all actors. New tools are needed to implement this approach, which focus on the process rather than the product: the result can be a city with a greater variety of uses, forms and time horizons [12].

The examples presented in this volume, through single lemmas or through in-depth essays, testify how the world of the temporary is a present and active reality in the city and can coexist with the elements of permanence, integrating itself among those fragments that are part of the city's collective memory.

4. The temporary nature of the relationship with citizenship that has variously stressed and distorted its meaning, according to the articulations that the present work seeks to investigate, detects of the centrality that the temporal dimension as such has assumed in our epoch. Among the dramatic changes that have eroded the fabric of society and collective life in recent years is a fundamental shift in the perception and use of time. The salient feature of modernity consists from this point of view in the production of the constant acceleration of social life and, concretely, the rapid transformation of the material, social and spiritual world [13]. Following Hartmut Rosa's argument, alongside a technological acceleration, which compresses and tends to annihilate the perception of space, a social acceleration is given, which concerns the change in the very rhythm of changes in society itself. The relationship between space of experience and horizon of expectations tends, as Reinhart Koselleck had already established [14], to make them coincide while the constant acceleration of the rhythms of cultural and social innovation produces a kind of "contraction of the present" [15]. From the latter thus results a growth in the rhythms of decay in the reliability of experiences and expectations and from the contraction of the time arcs definable as "present." Typically, such a process manifests itself in late modernity on the empirical level in areas such as family life cycles that tend to last less than the lives of individuals or the world of labor and employment where within the individual life span people increasingly change occupation and the "career" tends to fragment in a series of stops and starts in different directions, including dramatic joblessness, to the point of losing all reference to a unified and progressive course [16]. Which develops a

further acceleration of the pace of life, an exponential increase of actions in the unit of time, as Georg Simmel already saw in his *Philosophy of Money* [17]. The constant pressure to regard time as a scarce resource has been accentuated by phenomena such as the globalization process and the digital revolution, which instead of making time-saving available tend to demand an increasing number of actions and experiences within the same amount of time.

The reflection on domination and subjugation has also, in turn, placed the temporal dimension at the center of the pathologies it identifies and the possibilities for reversing course [18]. Having ascertained the epochal diagnosis that signals a strong discontinuity of the symbolic continents on which our common thinking and feeling rests, Remo Bodei's analysis focuses on the transformations undergone by the present, past and future, memory, and, especially in tune with Rosa, by the relationship between work time and life time. In a context of heightened social inequalities and a widening gap between the times of machines and the times of human beings, individuals are under increasing pressure to focus on the present, with its elusive and ever-renewing nature, to take refuge in a consoling and unchanging past, or to get sucked into a jumble of rapidly overlapping and disappearing future projects. The by definition temporary character of the present, which refracts in a phantasmagoria of thoughts images and feelings and resolves itself into a constant haemorrhage of energy, tends to become, however, the predominant dimension of experience, breaking the vital articulation between the three times and making the whole existence precarious. With a neurotized replication of the motions of the extractive and robbery economy, individuals are driven to seize every opportunity that comes their way out of fear that it will never come again, and to immerse themselves in a punctual present that also invades other temporal dimensions. The difficulty in identifying the contours of the future retroacts on the present by producing fear and demotivation. Individual consciousness retains the past less and is losing the ability to unwind it to re-join the present. Downloading into objectified memory devices the whole full of unselected memories becomes mostly an exoneration of individual memory and weakens forms of shared, social and collective memory.

Beyond the practices of good use of time left to individual conduct of life, there does not seem to be on the horizon any reconstitution of common and shared forms of reaction to the deep movement of the epoch as it has been described so far, except perhaps with regard to the climate crisis, albeit in a manner for the moment inadequate to the gravity of the problem. Regarding the aspect of temporal acceleration highlighted here, an attempt will be made to discuss the point that sees the possible reconstruction of a responsive relationship between us and the world through resonant forms or even through dissonant forms of relationship with the world of things, the social world, nature, and work [19].

5. The activities that gave rise to the volume presented here began with a survey of citizen studies and the current status of citizenship. With the intervention of Aoileann Ní Mhurchú (University of Manchester) on how citizenship is practiced in (un)familiar ways, the exploration of Critical Citizenship Studies literature began, in order to challenge us to move beyond what we know about citizenship—pushing us to consider actions which disrupt and rupture ideas about inequality and exclusion rather than reaffirm these. This survey considered how citizenship is practiced in unfamiliar ways through cultural acts which do not manifest as struggles over inclusion and exclusion in ways we have come to expect—through a familiar declaration along the lines of a statement which says "here is how we can see a challenge to the designation of exclusion". Instead, it explored how cultural acts indirectly affect the categories through which the nation-state operates—its practices of racialization, its cultural, sexual and social boundaries etc.—through the sensory processes they invoke (feelings, embodiments, soundscapes) and the importance of ambiguity

here. Thus, the focus is on to contribute to insights in Critical Citizenship Studies about the hybrid and ambiguous ways citizenship is lived, experienced, challenged and (re)configured, with a special interest in how aesthetic forms of meaning and representation expressed in language and music help us rethink understanding of "resistance", "voice" and "political subjectivity".

Gregor Fitzi's (Europa-Universität Viadrina, Frankfurt an der Oder) contribution on the crisis of citizenship as a symptom of societal destructuration, which also opens the present collection, proposed a framework reconstruction of the subject "citizenship" and its recent developments. Following his reconstruction, the deregulation that characterized the neo-liberal era since the 1980s led to a factual if not formal weakening of citizenship rights. In particular, social rights have undergone a major contraction, mainly due to the changes in labor market praxes at the limits of current legislation. Processes of dualization, precarisation and conditionality in industrial relations led to a progressive disconnection between official labor legislation and the factual reality of work, which increasingly orients itself towards provisional regulatory regimes that erode the validity of formally established legal frames of action. The result is a phenomenon of regulatory intermittence that needs to be reconstructed in sociological terms. The resulting "normative intermittency" that erodes access to citizenship rights can be seen as the symptom of a broader transformation of complex societies that is linked to ongoing processes of social de-structuration. These are characterized by a regression in normative social differentiation that undermines the social legitimacy of post-WWII welfare regimes and substitutes them with provisional modalities of "asymmetrical social consensus". This contribution develops an analytical framework to evaluate causes and effects of this societal transformation and to understand how it undermines the factual validity of citizenship regimes.

With interventions by Federico Rahola (University of Genoa) and Pierpaolo Ascari (University of Bologna), the book focuses on the places and forms of citizenship at the edge that escapes and demands recognition in transient cities and camps of gathering and transit. Confronting with the current proliferation of administrative detention facilities for displaced and illegalized persons, Rahola's contribution traces the different manifestations of this specific border apparatus back to a more general and abstract "camp form"—one whose origins date back to the colonial realm, finding in the colonial subject the first internable figure. To such institutional form, it opposes the relentless production of informal, occupied, and often clandestine encampments, dwelled by people on the move and scattered along as many hidden and illegalized routes. By rereading current makeshift camps and hidden routes though the historical lens of the US pre-civil war Underground Railroad, the article suggests to conceive of them in terms of as many counter-spaces of a possible "Underground Europe" whose material and unauthorized existence, often supported by criminalized solidarity networks, mirrors and reverses that one of a "Borderland" or "fortress Europe". By the same token, it suggests to conceive of these temporary, precarious and informal zones in terms of as many "reverse shot" of the institutional "camp-form", defining them as counter-camps and focusing on the political, spatial and temporal relation between these two opposed polarities.

Ascari's text, on his part, reconstructs the story of the transition that took place when the French authorities resolved to dismantle the Parisian bidonville, the new "dispositif" they resorted to was that of transit cities, a programmatically provisional solution that still today, however, awaits to be completely overcome. In Nanterre, for example, the definitive reconversion of the "cité des Potagers" into a district of rent-controlled housing will only be completed in 2025, but in the meantime the entire history of these complexes will have resulted in the chronicization of an "unfinished urbanity", as Abdelmalek Sayad defined it, determined on the border between the dimension of the temporary and the more implicit one of the class, gender and race hierarchies that the reference to the temporary allows to eternalize.

Another group of presentations and texts foregrounds the housing, architectural and urban dimensions of temporariness and citizenship. In this direction, Annalisa Trentin (University of Bologna) discusses how permanence and temporariness are two counterposed terms that characterize architecture and urban transformations. Her essay intends to address some of the many interpretations of this binomial: from the role of urban permanence and the temporaneity of use, as addressed by Aldo Rossi in *The Architecture of the City*, to the concept of the temporary duration of the life of buildings and the concept of conservation, presented by Rem Koolhaas with his research *Cronocaos*. Next, the fundamental role of unplanned areas is analyzed as a place for experimentation of the temporary, as a means to test future scenarios, taking as an example the survey conducted by the Urban Catalyst group, and then concluding with the experience of the IPA—Institute for Public Architecture, in the desire to reduce the transitory nature of living, avoiding gentrification and offering a permanent living condition to the most disadvantaged social groups. The examples presented are intended to illustrate how architecture and the city often live in the balance between permanence and transience, that there are no absolute instances and that only a well-structured architecture, on a formal basis that refers to a collective memory, can embrace the continuous transience of function and always remain vital within the urban organism.

According to Lidia Gasperoni (Technical University, Berlin), bubbles are concretizations of temporary spaces. On the one hand, the bubble constitutes a structural, inflatable element characterized by a specific, malleable, transparent materiality; on the other hand, it represents a conceptual spatial device for expressing a specific feature of contemporary anthropological spaces, a place of exposure, in which the boundaries between outside and inside, between soil and sky, seem to disappear. Her essay first reflects on the ephemeralization of space, focusing on the definition of space as a network. Second, it elaborates on this theoretical diagnosis by drawing on work by Vilém Flusser, who explicitly proposes the bubble as a space of contemporary living. Third, through discussing specific examples, it attempts to develop a critique of bubble spaces as a dimension of contemporary design.

In their talks, activist Monxo Lopez (Museum of the City of New York), and architect and historian Jonathan Kirschenfeld (New York) and Susanne Schindler (ETH Zurich), discussed the case of Mott Haven, a neighborhood located in the southern-most part of the Bronx, which has long been characterized by the temporariness of its residents. If substandard housing for recent immigrants was an issue in the 1930s, disinvestment caused the high turnover of the 1970s, and today, speculative real estate development results in involuntary displacement. Against this backdrop, Mott Haven has been a site for testing the construction of citizenship through housing. The City, often in conjunction with neighborhood-based initiatives, has employed a range of housing programs from public rental housing to nonprofit cooperatives to individual homeownership in an attempt to fight poverty and stabilize residents. This leads to the contradictory situation in which those experiencing high poverty and living in public housing are the ones most stable, but often the least active in local politics; while those who might be able to afford and be interested in alternate housing forms, including through Community Land Trusts, and are actively organizing to that end, are often rapidly displaced. The talks considered the tensions between temporariness and permanence, city and citizenship through the lens of housing in Mott Haven. They take the cue from an unrealized co-housing proposal that all three practitioners worked on cooperatively in 2019, looking back at the intentions and realities, as well as the legacies, of the Model Cities program (1966–74) and vest-pocket housing, as well as reviewing the underlying issues contributing to inequality in housing as related to speculative real estate development and land ownership. The Community Land Trust has been suggested as an alternative ownership model structure that can produce permanently affordable housing in low- and moderate- income communities. Finally, examples of his publicly funded housing projects for formerly homeless individuals (built by not-for-profit developers) within 'leftover' sites in Brooklyn and the Bronx were shown, mixing small studio units with on-site social services and congregate spaces and are termed "supportive housing".

Martina Massari, Arshia Eghbali and Serena Pagliula (all three University of Bologna) deal in their contributions, respectively, with transient communities for urban preparedness, the art of temporary living looking inside student rooms, and energy community and citizenship as enabling actions for integrated energy plan implementation and urban energy transition. Massari describes the way the crisis has been challenging cities and urban planners for decades, producing theories, scenarios, and imaginaries aimed at governing its emergence and its consequences. The latest of such crises—the Covid-19 pandemic—seems to have once again brought attention to the unsolved urgency of planning to tackle the changes and stresses caused by insurgent events, which once again appears to involve cities and their spaces as holders of both the reasons and the possible solutions to the crisis consequences. Despite the abundance of possible approaches, however, the debate has not yet clearly highlighted the operative lessons learned from these challenges. In this vein, the urban studies debate has been reflecting on the possibility to act provisionally but in a preparedness perspective, opening alternative paths, rather than proposing solutions, planning for uncertainty and complexity with temporary means and actions. An attitude that requires both the flexibility and adaptability of consolidated urban systems and the affirmation and legitimization of collective and practical instances in an operational institutional dimension. The article briefly reviews the relevant positions in the relationship between cities and crisis; next, it highlights the role, responsibilities and relevance of planning to inhabit the consequences of the crisis, from a preparedness perspective; lastly, it calls to consider the potential alternative answers resulting from interaction with temporary communities of practice.

Through four brief fragments, Eghbali establishes a cause, an approach, and a context for an aesthetic inquiry into student rooms as places of transitory dwelling and materializations of an art of temporary living. First, based on the notion of the "arts of doing" (*arts de faire*), the art of temporary living is defined as the ensemble of tactical, everyday practices that shape and are shaped by the students' landscape of home. Second, the infra-ordinary gaze is presented as an approach for investigating how the basic level of everyday life is lived through its mundane practices and materialities, Third, through a discussion of Georges Perec's novel *La Vie mode d'emploi*, and by situating it within the wider context of the visual culture of voyeuristic illustrations of urban domestic interiors, a certain aesthetic is defined for such illustrations along the lines of architectural precision coupled with an exhaustive attention to the mundane traces of lived space. Fourth, the student bedroom is discussed and studied through original illustrations created based on ads posted on online platforms for renting or sub-letting rooms to students in Bologna, Italy. Finally, two recurring material elements of the student room are explored in order to shed some more light on the art of temporary living as practiced by university students.

In her paper, Pagliula suggests that one of the main challenges of the future is to create climate-neutral and smart cities, with zero CO_2 emissions by 2050. Cities have to be conceived, planned and designed in order to remain competitive and survive growing populations, scarce resources and changeable and unpredictable built environments by putting sustainability, health and the quality of citizens' lives at the center. The contribution aims to investigate the link between integrated energy plans developed by cities and the emergence of energy community as enabling action for the implementation of plans, highlighting how integrated energy plans can favor the emergence of energy community, as legal entity, and energy citizenship, as new definition of consumer, and help to move from the planning to implementation phases, as well as opening to new questions that will be developed in future research.

Before the Lemmary of Temporary Citizenship, to which the doctoral students who participated in the entire research pathway contributed in order to investigate the notion of citizenship in a preliminary dimension to the realms of its formal definition, the book presents three insights dedicated to images of the people as *populus*, *plethos*, *plebs* and *ethnos* (Jacopo Galimberti, IUAV, formerly University of Bologna), to inhabiting the ecological conversion (Andrea Mubi Brighenti, University of Trento) and to reflecting on dissonant temporality as constitutive of the contemporary experience of time (Andrea Borsari, University of Bologna). Galimberti notices that numerous scholars have explored the genealogy of concepts such as "citizen", "the people" and "sovereignty" from the perspective of the history of ideas and political sciences. In the last fifteen years, however, the academic interest in these analyses has been supplemented by lively debates that have occasionally reached the mainstream public arena in the Global North. In this context, social media platforms have played a major role, circulating innovative "images of the people" that tap in the history of political iconography. The essay focuses on the people as a performative entity whose contemporary visual instantiations deserve theoretical analysis. In particular, it combines the conceptual tools of political iconography with the framework delineated by a recent study about four meanings of the term "the people" (the people as "plebs", as "populus", as "ethnos" and as "plethos") in order to analyze brief case studies, and to suggest the need to conduct more in-depth analysis of visual negotiations of "the people" on social media. For his part, Mubi Brighenti questions the issue of ecological transition from a political perspective. An attempt is made here to reassess Alex Langer's approach to ecology through the lens of "conversion", examining the implications of his legacy for a new vision of citizenship. A suggestion is made to consider what could be the alternative to a "symbolic" take on the ecological transition, retrieving a few indications from Peirce's pragmaticist philosophy.

In conclusion, Borsari aims at analyzing the different layers of meanings contained in the notion of "dissonance" from the aesthetic, cognitive and emotional point of view. To then make them available for the exploration of the different forms of "dissonant time" analyzed by contemporary thought, in particular by thinkers such as H. Blumenberg, R. Bodei, R. Koselleck, E. Traverso, E. Bloch, H. Rosa, G. Wilder, field studies of "new urban worlds" and political philosophical approach to concrete utopianism. In this context, notions such as the following become relevant for understanding the temporal experience of late modernity: gap between world time and life time; intertwining and overlapping of future-facing past and present; changing and at different speeds "geologically" multi-layered relations between space of experience and horizon of expectation; "presentism" and breakdown of social frameworks of memory and individualistic reification of the past; "contemporaneity of the noncontemporary"; montage and kaleidoscopic temporal dialectic; disconnection between spheres of economic, social and generational life and lack of time; multiple temporalities and cross-connections across urban spaces, as well as, plurality of times in a political perspective of concrete utopianism. So that dissonance seems to be the most appropriate conceptual and figural tool to effectively render the condition of contemporary temporal experience in its real irreconcilabilities and perspectives.

The explorations in the form of headwords for key terms of the various implications of the notion of temporarity in relation to citizenship, architecture, and the city, conducted by doctoral students are complementing the book: Andrea Cattabriga (Assemblages/Decentralised), Valentina De Matteo (Transition/Deviance), Francesco Di Maio (Hacktivism/Sexual tourism), Lorna Dragonetti (Life cycle thinking/Life cycle phase), Clara Giardina (Time-based packaged goods/ Luggage), Marco Iannantuono (Adaptation/ Cueva and Tomas de terreno), Giulia Marzani (Temporary uses/Pop-up spaces), Angelo Massafra (Classroom/Digital Citizenship/Material Culture), Claudia Nigrelli (Grenze vs. Schranke/ Open form), Marco Palma (Deliberation/Climate Justice), Dafni Kalliopi Retzepi (Border/Rules).

References

1. BABELS: Entre accueil et rejet: Ce que les villes font aux migrant, le passager clandestin, Paris (2018)
2. Isin, E.F.: Performative citizenship. In: Shachar, A., Bauböck, R., Bloemraad, I., Vink, M. (eds.) The Oxford Handbook of Citizenship, Oxford University Press, Oxford (2017)
3. von Uexküll, J.: Ambienti animali e ambienti umani. Quodlibet, Macerata (2010)
4. Guillaumin, C.: Race et nature: système des marques, idée de groupe naturel et rapports sociaux. Pluriel (11), 39–55 (1977)
5. MacCannel, D.: The Tourist: A New Theory of the Leisure Class, pp. 110–133. Schocken Books Inc., New York (1976)
6. Foucault, M.: Leçons sur la volonté de savoir. Cours au Collège de France (1970–1971) suivi de Le savoir d'Œdipe, Seuil/Gallimard, Paris (2011)
7. Foucault, M.: Teorie e istituzioni penali. Corso al Collège de France (1971–1972). Feltrinelli, Milano (2019)
8. Pratt, M.L.: Arts of the Contact Zone. Profession (1991)
9. Bishop, P., Williams, L.: The temporary city, p. 19. Routledge, London New York (2012)
10. M.D.A. Rounsevell: Identifying uncertainties in scenarios and models of socio-ecological systems in support of decision-making. In: One Earth, vol. 4, issue 7, pp. 967–985, July 23 (2021)
11. Sendra, P., Sennet, R.: Designing Disorder. Experiments and Disruptions in the City, p. 127. Verso London, New York (2020)
12. Giseke U.: Stadtumbau und Freiflächenentwicklung: Zwischennutzung und neue Freiflächen – städtische Lebensräume der Zukunft, lecture presented at Difu Seminar, 'Neue Brachen und Flächenpotenziale: Nutzungsmanagement als kommunale Herauforderung', Berlin, 13–15 June 2005. In: Blumner N.: Planning for the Unplanned: Tools and Techniques for Interim Use in Germany and the United States, Deutsches Institut für Urbanistik, Berlin (2006)
13. Rosa, H.: Alienation and Acceleration: Towards a Critical Theory of Late Modernity. NSU Press, Natchitoches-LA (2010)
14. Koselleck, R.: Futures Past: On the Semantics of Historical Time. Columbia University Press, New York (2004 [1979])
15. Lübbe, H.: The contraction of the present. In: Rosa, H., Scheuerman, W. (eds.) High-Speed Society. Social Acceleration, Power and Modernity, pp. 159–78. Penn State University Press, University Park-PA (2009)
16. Sennett, R.: The Corrosion of Character: The Personal Consequences of Work in the New Capitalism. Norton, London (1998)
17. Simmel, G.: The Philosophy of Money. Routledge, London (2011 [1900])
18. Bodei, R.: Dominio e sottomissione. Schiavi, animali, macchine, Intelligenza Artificiale [Domination and Subjugation. Slaves, Animals, Machines, Artificial Intelligence.]. Il Mulino, Bologna (2019)
19. Rosa, H.: Resonance. A Sociology of Our Relationship to the World. Polity, Cambridge (2019)

Pierpaolo Ascari teaches Aesthetics at the Department of Architecture, University of Bologna. He is author of *Ebola e le forme* (Rome, 2017), *Attraverso i confini. Lettura, storia ed esperienza estetica in Stendhal e Flaubert* (Milan, 2018), *Corpi e recinti. Estetica ed economia politica del decoro* (Verona, 2019) and *The Adventure of Form: Aesthetics, Nature and Society* (Delft, 2021). He edited the volumes *Oggetti contesi. Le cose nella migrazione* (Milan, 2020), *Espressioni urbane. Muri sconciati, writing e street art* (Milan, 2022, whit Pietro Rivasi) and *I luoghi delle merci* (Milan, 2022).

Andrea Borsari teaches Aesthetics at the Department of Architecture of the University of Bologna, directs the series "La vita e le forme" (Bup) and is deputy director of "Iride. Rivista di filosofia e discussione pubblica" (Il Mulino). He was recently visiting professor at UCLA, ENSA-Paris La Villette, the 4 Cities Unica Euromaster in Urban Studies—University of Copenhagen and the University of Sichuan, and visiting fellow at University of Amsterdam, University of Buenos Aires and Brown University—Providence. Among his publications: *Mondo, cose, immagini. Sulle forme dell'esperienza estetica* (Bup, 2018), *Mimicry. Estetica del divenire animale* (Mimesis, 2018) and, as editor, *Philosophical Anthropology and Critical Theory* (Iride, 02/2022), *Aesthetics of the Anthropocene* (CPCL, 1-2/2022, with P. Ascari).

Annalisa Trentin is an architect and full professor in the field of architectural and urban design at the Department of Architecture of the Alma Mater Studiorum—University of Bologna; since 2018 she has been Coordinator of the Ph.D. course in Architecture and Design Cultures (Unibo); from 2015 to 2021 she held the position of Deputy Director of the Department of Architecture, where she has been carrying out research and interdisciplinary teaching activities since 2001. She carries out design research by participating in international workshops and competitions; she is the author of several essays on pedagogical practice in architectural design, criticism and theory of contemporary architecture.

Exploring Temporary

The Crisis of Citizenship. A Symptom of Societal Destructuration

Gregor Fitzi

Abstract

A widespread socio-political diagnosis suggests that renewed reform processes are needed to address the social and environmental crises facing contemporary societies. To understand what this implies from the viewpoint of societal analysis, a retrospective look at the reform processes that characterised the substantial development of welfare after the World War II as well as its later crisis is necessary. The chapter reconstructs these processes with reference to the United Kingdom during the Clement Attlee's Labour Government since 1945. It shows how the institutionalisation of civil, political and social citizenship rights influenced the building of a shared consensus about the fundamentals of civil coexistence. To understand its later crisis the guiding principles of the neo-liberal worldview as well as the modalities of their implementation since the 1980s are presented. The focus is here on the social fragmentation and the loss of social legitimacy that these processes induced. The main societal phenomenon that emerges from the analysis is that of a failing social structuration that comes to expression in the 'intermittent normativity' of contemporary societies. The question thus arises as to how a resumption of regulatory societal structuration in the sense of an ecologically aware twenty-first century welfare can develop. In this respect, the chapter analyses the reciprocal action between the high social plasticity of contemporary societies and the transformative potential of social action.

Keywords

Citizenship rights · Failing social structuration · Intermittent normativity · Societal plasticity · Social legitimation · Transformative social action

1 Introduction

The institutionalisation of citizenship rights, especially in their social component, is the result of a specific social history that has shaped post-war societies in Europe [1]. In its different variants, the era of the European welfare state has brought about both an improvement in the living conditions of the subaltern classes and robust economic growth [2]. The sociological reflection on this development favoured the inquiry into 'citizenship status' as the crucial institution granting the social integration of complex societies [3]. However, the oil crisis and stagflation during the 1970s pushed the European welfare

G. Fitzi (✉)
Faculty for Social and Cultural Sciences, European University Viadrina, Frankfurt (Oder), Germany
e-mail: fitzigregor@gmail.com

A. Borsari et al. (eds.), *TEMPORARY: Citizenship, Architecture and City*, The City Project 4,
https://doi.org/10.1007/978-3-031-36667-3_2

models into crisis. This could have been addressed through an ecological reconversion of the production system and an adaptation of social protection to the new challenges of increased international competition, deindustrialisation, and demographic change. Yet, ever since, societal transformation has been led by an ideological bias that has pursued the goal of dismantling the embeddedness of the economy in society.

The unfolding of the neoliberal revolution since the 1980s has massively eroded the social backdrop of citizenship rights, thereby also weakening civil and political rights of subaltern classes [4]. In an ideal–typical way, this transformation can be reconstructed by studying the progressive dismantling of labour market protection legislation. This inquiry is key to understanding the characteristics and scope of the ongoing societal destructuration. Yet, in the process, the analytical tools of sociology must be fundamentally renewed [5]. The ongoing dynamics of societal fragmentation must be studied in an innovative and thorough manner, because it implies phenomena like 'normative intermittency' and 'uncontrolled societal plasticity'. We can only ask about the possible ways out of the ongoing 'failing societal structuration' if the social sciences get on the track of the continuous alternation between processes of social differentiation and destructuration that characterise complex societies. Yet, answering this question implies to understand how 'transformative social action' can lead from the predominance of temporary citizenship forms to the restoration of its full institutional range and potential.

2 Institutionalisation of Citizenship Rights: Civil, Political and Social

Under the leadership of Clement Attlee's Labour Government, the post-war 'Welfare State' was built on the awareness of the worst consequences of capitalist production cycles, integrating important measures of state intervention in the economy [6]. This societal structuration process, which started with the Labour Party's surprise landslide victory in the July 1945 elections, became a model of comparison for the different reform processes that started later around continental Europe, including the German 'social market economy' that developed instead in the spirit of Ordoliberalism [7]. The establishing of the welfare state in the UK thus has an ideal–typical character [8], which strongly influenced the sociological reflection on the 'normative status' of citizens that is necessary to protect the social strata most exposed to the cyclical crises of capitalist accumulation processes.

Labour's welfare-state legislation of 1945 could build on the preparatory work of the so-called 'Beveridge Committee' that was set up by the all-party coalition government during the war to understand the measures to be taken for post-war reconstruction [9]. Besides the sober assessment of facts and figures, the committee's report exposes the normative principles that justify the vision of society, which a reform and improvement of social protection services was supposed to realise. Three main principles were put forward ([9]: 6). These were to determine the normative orientation of the legislation to be adopted, so that the organisation of social insurance could be treated as part of a comprehensive 'policy of progress' (ibid.). Its first goal was to provide 'income security', which above all required 'an attack upon want' (ibid.). Yet, the realisation of the necessary measures implied a reciprocal normative commitment of the community and the individual. This meant that the state should grant minimal social protection and individuals take care of the further needs of themselves and their families. Thus, both were set a common goal: to move forward from an era in which deprivation represented a tolerated social condition.

The principles that the *Beveridge Report* put forward were not specifically socialist, yet they testified to an acknowledgement of the demand for limiting the impact of income discontinuity, even if they underlined the idea of an obligation of the state to guarantee only the minimum necessary for survival. In any case, the idea that society should grant income continuity for everybody, because of the discontinuity of capitalist production cycles, had established itself as a shared normative principle, even within the circles of liberal economists. The achievements of the Attlee government in translating the idea of the welfare-state compromise into legislation established the so-called post-war consensus [10]. The cultural hegemony of the welfare-state compromise lasted during the following three decades and was only dissolved with Margaret Thatcher's accession to power in 1979 that saw the neoliberal era ushered in. The implementation of the post-war consensus in different European countries initiated a social change that substantially improved the living conditions of the most disadvantaged social strata. This regulated transformation process changed European societies from predominantly agrarian to industrial with a highly developed tertiary sector, it realised a great leap forward in education and, last but not least, provided the economy with framework conditions that enabled three decades of growth. The consequences that this development had on people's everyday lives is shown exemplarily by Fourastié's reconstruction of the French case in the book whose title became a catchphrase—'The Glorious Thirty' [2].

Sociological reflection quickly raised awareness about the extent of the institutional and cultural, as well as economic, transformation that was taking place in the wake of the welfare state compromise, as Thomas H. Marshall put forward already in 1949 [3]. He underlines how in the UK by the end of the nineteenth century a tradition of economic thought had developed, which started from liberal positions and became receptive to socialist claims. The sociological assessment of this heritage starts from a consideration on human dignity—that is, from the assumption that each human being has the potential and thus the right to fully develop all his capabilities, independently of his material living condition. In the face of this finding, the distinguishing features of modern capitalism were the uncertainty of life conditions and the burden of heavy and excessive labour that prevented working classes from achieving the goal of self-realisation, so that the framework conditions of their lives had to be changed.

A central point of reflection here is the issue of how the existence of the market economy can be integrated with the self-realisation of every human being. The direction of reflection thus moves away from the pure quantitative assessment of the standards of living in terms of goods consumed and services enjoyed towards a qualitative assessment of life in terms of essential elements in civilisation and culture. Marshall raises the question of the parameters granting effective membership in human society, which not only implies the juridical formulation, but also a substantial access to specific rights. On this path of reasoning, citizenship sociology encounters its crucial social structural normative issue. Legal concepts of full membership in a community—that is, of citizenship—struggle to become consistent with the inequalities, which conform to the economic stratification of society.

Hence, the discussion in principle is enriched with means of empirical evaluation that allow to understand to what degree effective membership is granted in society. Historically, the basic membership status in a community has been enriched with new substance, by adding to it a formidable array of three sorts of citizenship rights: civil, political and social. Yet, this societal transformation could not be achieved without intervening with strong regulatory measures into the uncontrolled freedom of competitive markets. The logics of social democratic market regulation represent the means to realise citizenship in a comprehensive sense. For Marshall, however, this process of societal structuration is based on a conflict of principles which demands examination.

Citizenship is a status bestowed on those who are full members of a community. All those who possess the status are equal with respect to the rights and duties with which the status is endowed. There is no universal principle that determines in detail what those rights and duties shall be. Yet, societies in which citizenship becomes an institution orient themselves to an ideal of citizenship against which achievements can be measured and towards which aspirations can be directed. Social class, on the other hand, is a system of inequality. And it too, like citizenship, is based on a set of ideals, beliefs and values, which instead foster exclusion.

It is therefore reasonable to expect that the impact of citizenship on social class structuration takes the form of a conflict between opposing principles. The crucial question of citizenship sociology—that is, of the welfare-state compromise—is thus how the opposing principles of membership in a community and belonging to a socio-economic class can grow and flourish side by side in the same soil. With the post-war consensus, this achievement seemed to become possible. Later, however, it lost legitimacy and was consistently eroded.

3 Neoliberal Revolution

The normative conception of a welfare-state compromise for complex societies, which was implemented through the pressure exerted by trade unions and political parties after WWII, has not survived the attacks of the neoliberal revolution. What Marshall called 'social class' has been widely fragmented and results today in a composite phenomenology of conflicts, whose understanding is much more difficult. Discriminated social groups composed of women, precarious workers, migrants and outcasts, who are not or only partially taken care of by social protection systems, constitute the members of society, whose membership is not recognised in a substantial and sometimes also in a formal way. These groups are characterised by socio-cultural features of a complexity that cannot be reduced to the person's economic position in society, in the sense of Weber's *Klassenlage* (MWG I/23 592–595) [11].

Accordingly, societal self-interpretation classifies the related conflicts as gender discrimination, lifestyle intolerance, and cultural and ethno-religious tensions. The common traits of these social phenomena remain underplayed, so that the overall social change induced by today's 'crumbling late capitalism' is not sufficiently investigated ([5], 27–40). Here, sociological research can contribute to regain a comprehensive understanding of the ongoing societal restructuration by showing which common traits of exploitation, exclusion and negated access to citizenship rights characterise the different groups. This reflexive work beyond the classifications produced by societal self-interpretation may deliver a knowledge basis for an up-to-date reformulation of the welfare state compromise. On this basis, a renewed process of normative societal structuration could take place. While the awareness of this necessity seems to have been revived after the global pandemic of 2020/21, the existing societal conditions are as distant as ever from this possibility.

The limits to improving social equality, which the welfare state systems already showed before their crisis, were magnified by the policies adopted in the wake of the neoliberal revolution in progress since the 1980s. Piketty summarises the transformation in the following points ([12]: Chap. 11). A progressive withdrawal of public institutions from ownership of strategic industries, the dismantling of progressive taxation, and a weakening of co-determination schemes between stakeholders and workers' representatives has resulted in a renewed growth of income inequality in Europe, the United States, China and India. Added to this development are the massive deregulation policies that set out the frame conditions for the financial and economic crises of the years ahead [13].

Basically, the neoliberal worldview postulates three principles: (1) Real existing global capitalism knows no alternatives (Thatcher's famous dictum). (2) Its model of development does not tend to crisis, but rather undergoes 'reasonable adjustments'. (3) Consequently, any political regulation of the economy that is not dictated by economic interests is inadmissible. Despite its ideological leanings, neoliberalism has never refrained from letting the state intervene in a regulatory way in economy and society, yet only for the benefit of capital investment. As a corollary of the political principles of the neoliberal worldview, who or what bears the social and ecological costs of the repeated 'readjustments' that characterise the development of real-existing global capitalism is considered irrelevant. This claim is linked to a completely uncritical faith in the ability of the market to regulate not only itself, but the whole of society, so that in accordance with Adam Smith, time and again, it is supposed to find a new balance by an 'invisible hand' and not require any regulation [14].

Neoliberalism entered the stage of media-oriented politics with the promise that, thanks to the globalisation of uncontrolled capital investment, so much wealth would be released to generously compensate for its social and ecological costs. The belief in the miraculous power of a 'linear development of the productive forces' thus still formed the basis of the ideological turn of the 1980s. In such an atmosphere, many politicians and intellectuals were seized as if by a spirit of religious conversion, driven by a non-reflected faith in the historical power of a zeitgeist that, in a Hegelian sense, had spoken with the failure of real existing socialism. Thanks to this 'social effervescence', the neoliberal creed was able to make inroads so easily into the political culture of the left and turn many social democrats and post-communists into ardent supporters of the globalist utopia [15, 16]. They naively expected that an unregulated global economy would rapidly achieve maximum development of the productive forces, therefore leading 'with scientific certainty' to a post-capitalist age, or at least providing a cornucopia of resources for redistribution.

Yet, the redistribution of globally produced wealth, in whose favour the welfare state legislation was adopted since the New Deal, was by no means the objective of the neoliberal project. These measures together with the regulation of the financial markets were to be dismantled [17]. After the fall of the Berlin Wall, when the left was still dreaming of the prospects of democratic socialism beyond the mortgages of Stalinist dictatorship, the architects of the new global economic order were rapidly pursuing their political project. They knew that time was short to impose the neoliberal revolution before societies would fight back [4]. So they interpreted the dissolution of the Soviet Union as a historic signal to finally overturn the entire regulation of capital and labour markets in both East and West. The aim was to impose an uncontrolled rollout of capital investment that went far beyond what the pioneers of neoliberalism had achieved in the 1980s. Any restriction to the limits of capital accumulation to protect society and the environment was to be blown up, and not only the real-socialist variant of the political control of the economy.

The ideological underpinning for the political strategy of neoliberalism was given with the overturning of the social and ethical concept of 'responsibility'. This allowed the inversion of the normative hierarchy between the political logic of the common good and the economic logic of material interest. The miners being laid off, as well as the unemployed in general, were relied upon to embrace their individual responsibility for the predicament that hit them. With a revival of the Calvinist doctrine of predestination, the condition in which they found themselves was interpreted as the product of their own shortcomings (MWG I/18), thus absolving the economic development model as well as its democratic control of any responsibility. The 'new spirit of capitalism' thus extended to the workforce which was forced to rediscover its ascetic vocation [18]. The normative principle of the welfare state compromise was denied, according to which society as a whole is responsible for those affected by the consequences of the irrational development and crisis dynamics that characterise the capitalist system of production. Yet, this 'reversal of values', founded ideologically on the principle of responsibility, did not affect everyone equally, because at the same time, in the neoliberal age, large-scale bail-out operations were repeatedly launched to rescue financial institutions from the bankruptcies caused by their reckless speculation.

Since the 1980s neoliberal revolution, the 'arrangement of society' has thus acquired a new structure. To the exclusive advantage of economic actors, the economy and politics stay in a relationship of hierarchical structuration that has specific consequences. On the one hand, the social and environmental costs for economic mismanagement and the reckless financial speculation of private investors are transferred to society as a whole. Contrary to the promises of the globalisation utopia, the predominant mechanism of neoliberal policies entails the characteristic that profits of accumulation processes are not socialised, while their losses are very well socialised. On the other hand, at every new turn of the crisis the remorseful representatives of business and politics announce new legislation to avoid such excesses in the future. Yet, when the emergence phase ends, no specific measures are adopted, so that the uncontrolled speculative dynamic starts again from the beginning.

4 Social Fragmentation

The social costs of irrational societal development under the neoliberal regime can be exemplified with an analysis of the increasing fragmentation of labour markets. It shows how the separation took place between the sectors of permanent employment, which are still covered by protective legislation, and the vast prairies of wage labour that have been conceded to open exploitation. The keyword here is 'dualisation' [19]—that is, in general the process by which society's insiders and outsiders are increasingly treated differently with respect to access to workers' rights ([20]: 57–75). In recent decades, the dualisation of labour markets has established, in particular, an institutionalised divide between the permanent and precarious workforce [21]. Changes in the structure of industrial relations, particularly the increasing number of precarious types of jobs, are converted into social and labour market policy measures, often in the form of exceptions to the overall legislation, so that provisional legal orders undermine the universalism of constitutional principles and social protection. The resulting normative conflict establishes the principle that social orders do not have a universal and necessary validity, but rather are intermittent in space and time—a condition that characterised pre-modern societies, as Weber's classical study on urban illegitimate orders shows (MWG I/22-5). This social condition comes back today in the form of an overarching 'normative intermittency' of complex societies [5].

Added to these phenomena is an increasing conditionality in the access to social protection measures that establishes requirements for people to behave in prescribed ways in order to access cash benefits or other welfare support. Conditionality is typically enforced through benefit sanctions of various kinds, reflecting a new vision of welfare, which is more focused on promoting disciplined behaviour than on protecting people against classic social risks like unemployment [22]. The social structural output of the interlaced processes of dualisation, precarisation and conditionality is a growing intermittency in access to citizenship rights, therefore agency must struggle to align with this normative inconsistency.

The transformation of societies disseminates the perception that the illegitimated factual order, often dictated by the changing economic imperatives of crumbling late capitalism, erodes the legitimated legal order, by producing 'normative dualisation'. The legislative framework of society is formally still in force, yet factually it is increasingly disregarded by a socio-economic praxis that induces growing inequality and discrimination between social groups ([20]: 39–55). An increasing number of society's members suffer a status-loss or get no full access to citizenship rights, so the legitimation basis for the rule of law narrows. The rest of society fears slipping into such a condition and becomes sensitised for political narratives that promise to restructure the welfare state on a restrictive ethno-racial basis.

Contrary to popular perception, these changes affect both the winning countries in the global economic challenge and those which are suffering the most, as similar developments can be observed in various welfare state typologies, for example, concerning zero-hour contracts [23, 24]. At the same time, they extend not only to marginal or outdated sectors of production but also to the leading sectors of large-scale industry like the automotive industry. After the end of the Fordist era in industrial production, four changes have left a lasting mark on wage labour: feminisation, flexibilisation, fragmentation and financialisation [25]. The overall transformation of production strategies has become increasingly important and has enforced a far-reaching fragmentation of labour relations. These do not primarily aim at in-house disaggregation of work, but at splitting off, redistributing and recomposing a multitude of work processes in and between different work organisations and organisational fields. As a result, the value chains are reorganised in production constellations, where very different labour and employment requirements coexist, even if they are directly dependent on each other ([26]: 139). The cooperation dynamics between workers and work units changes completely, as can be typically observed in the development of the automotive industry, where fragmentation tendencies are particularly evident.

Here, the single process sections (and the people working in them) are always 'in reserve' and find themselves just temporarily in the replaceable shell into which the factory has been transformed. Uncertainty thus becomes the keynote of industrial relations. For the time being, the individual has been able to prove himself as an efficient unit according to the applicable criteria. Yet, nobody knows what will happen tomorrow. Every arrangement is under permanent evaluation and labour contracts last, in the best case, for three years. All work is project-based within a limited time frame, so that the access to acknowledged social rights like full-time, regular labour contracts and the prospect of continuity in employment is questioned in a structural sense. The risks of the renewed global division of labour – above all its increasingly visible and pointed fragmentation—thus becomes evident to all actors and strongly limits workers' bargaining behaviour and trade union organisation.

The resulting dualisation of employment conditions, institutionalising different treatment for equal work, is perceived by workers as a failure of the overall labour market legislation, so they call for its reinstatement ([26]: 168). Within existing welfare state systems that legitimise social inequality through measures of social protection, growing islands of extra-legality establish themselves and thrive. Social fragmentation generates normative intermittency that is perceived as a failure of the formally stated legal orders. Claims for renewed normative structuration are formulated, yet collective social agency struggles to form itself and trigger processes of societal transformation in highly fragmented environments. Social fragmentation, however, cannot be simply classified as a side effect of current economic restructuration. It is rather the most visible aspect of an overarching societal transformation that can be observed in Europe's industrial heart [27] and extends to the countries of Southern Europe that become increasingly dependent on it [28].

In Germany, real net earnings and productivity substantially grew in parallel from the 1950s until the beginning of the 1990s. Yet, since 1993 the two curves have diverged; productivity has continued to grow, but annual real net earnings have decreased. Average wages (a figure that also includes the highest salaries) have now been falling for nearly twenty years. The resulting social divide is attested by the development of household income between 1998 and 2014. The upper 60% have recorded an increase in household income, whereas the lower 40% have experienced a significant decrease. The gap in living standards and income has widened as a result of a snowball effect leading to losses in healthcare, welfare state services and pension contributions, later causing old-age poverty and leading to intervention by the public sector to limit it. Phenomena that despaired in the golden age of the West German welfare state have reappeared, including the increase in occupational discontinuity, the spread of the employed, who are in poverty, and the fact that a stable orbit of employment, if ever, is reached only in an individual's forties or later.

The situation in Italy—as the ideal type of a declining industrial power [29]—is far more problematic when compared with the social fragmentation in a country like Germany, which belongs to the winners of globalisation. Here, recent waves of flexibilisation in labour market relations have easily combined with traditional trends toward illegal employment, so that the divide between the legal framework and the factual practices of wage labour have reached an unprecedented scope [30]. Fana's political critique of this development gives astonishing insights into the phenomenology of the Italian way to social fragmentation [31]. As in other countries, in recent decades several other phenomena have developed in Italy besides 'on-call' work, including the expansion of the gig economy and the misuse of students for internships, or civilian volunteers to replace missing permanent employees in companies, museums and public administrations ([31]: 57–90). The factual expansion of dual markets coincided with the measures intended to legalise marginal or illegal labour relationships, which existed alongside contractual work. However, in reality, this has led to a massive extension of precarious work, by substituting and fragmenting regular positions in hundreds of low-cost precarious tasks.

5 Legitimation Crisis

Back in 1973, Habermas formulated the question of what social crisis means from a sociological viewpoint as a matter of examining the 'systemic malaise of complex societies' [32]. Marx's publication of *The Capital* in 1867 heralded the sociological concept of social crisis [33] (MEGA II/5) which gained renewed currency for the analysis of the Great Depression of the 1930s. In Habermas' perspective, the concept addresses the persistent disturbance in the 'system integration' of qualitatively differentiated societies, whose underlying causes lie in the conflict between the contradictory imperatives of economy and politics. According to Marx, this societal imbalance is to be classified as the structural expression of the unequal distribution of socially produced wealth arising from private ownership of the economic means of production. In the historical phase of welfare-state regulation for capital accumulation since 1945, societies proved capable of controlling the systemic tensions of modern capitalism, by ensuring economic and social development [2]. Yet, the massive deregulation induced by the neoliberal revolution set in motion an uncontrolled crisis-dynamics [4]. One of the macroscopic symptoms of the continued disruption that affects system integration today is the ongoing process of colonisation, which leads to the steady subjugation of different social spheres to economic logic. By and large, every attempt has failed to limit the imperialist tendencies of the economic sphere and bring it back under normative criteria that facilitate society's common good. Contrary to what happened after the 1929 stock market crash, even as the damage of unbridled global capitalism became so obvious as to warrant fixing the consequences of the financial crisis in 2007/08 with a massive injection of taxpayers' money, no resilient regulatory framework was put in place to protect against its risks [34].

As Habermas pointed out in 1973, disruptions of 'system integration' can only be considered as threatening to the existence of societies to the extent that 'social integration' of societies is at stake, that is, when the basis of consensus for the normative structures of society is compromised to the extent that it rapidly loses legitimacy and social interaction becomes largely anomic. This diagnosis gets to the heart of the social and cultural crisis that affects the institutional framework of contemporary societies. Here, the prevailing socio-economic development model denies increasingly large social groups the spatiotemporal and social resources that they would need to engage in a reflexively-rational, let alone critical, evaluation of the logic of social structuration. Hence, the normative legitimation of society consistently weakens, establishing a *modus vivendi* between qualitative differentiated societal domains, which fosters the provisional acceptance of societal arrangements that are imposed by contingent economic and political imperatives.

The alteration in the processes of normative social structuration has anomic consequences, because social actors temporarily accept it by means of 'pragmatic consensus', without legitimising the compelling action frames imposed by the progressive colonisation of society through the economic logic. Accordingly, as Lockwood pointed out as early as 1964, the sociological concept of crisis must grasp the connection between systemic and social integration in its current historical specificity. Following

Habermas, sociological theory speaks of social integration when social actors are socialised in symbolically structured contexts of action that make them rationally-reflected bearers of normative social structures. However, if social integration suffers from a loss of legitimation caused by an increasing failure of normative structuration, the cohesion of society is fundamentally endangered.

In the structural-functionalist view of society that was still possible in the early 1970s, the focus was on the 'control aspect' that made society a consistent aggregate that might be steered by the normative framework of 'bureaucratic organised capitalism'. In this context, societies could be conceived as social systems capable of establishing a possibly conflictual, yet still coordinated overall logic that secured system integration. Here, institutional orders had to justify their claims to validity, by turning social actors into bearers of normative structuration processes that encouraged them to engage in critical-reflexive procedures of social integration. Today, social reality has radically changed shape. The dialectics that unfolds between the logic of social structuration and social action continues to mark the integration dynamics of complex societies. Yet, the crucial difference is that system integration has increasingly been replaced by colonisation processes between qualitatively differentiated spheres of society which fundamentally call into question the possibility of 'overall societal control'.

The disorientation of a completely deregulated model of development and the resulting intensified rhythm of societal transformation thus generate strong tensions and conflicts that any systemic logic can hardly capture. The result is a fundamental delegitimation of the established normative framework of society, so that social integration remains comprehensively undermined. Yet, the rapid and irrational socio-economic change does not lead to an abrupt collapse of society. The liquefaction trend of its institutional framework results instead in an increased recourse to asymmetric relationships of temporary consensus that prevent the resumption of normative structuration.

In the early 1970s, Habermas could still assume that the social systems based on welfare state measures were legitimated by means of normative justification, because they interpreted and acknowledged social needs. Today, even in leading industrialised countries an increasing number of social actors are confronted with the experience of having to support normative frames of social action, without these ever fulfilling their needs. In the age of welfare capitalism, above all in the German 'social market economy', social science could assume with a clear conscience that societies can manage their exchange with the natural environment through techno-scientific procedures and regain social balance thanks to the normative structuration potential of communicative action. Yet, after forty years of neoliberal societal restructuration, both aspects of 'social and environmental integration' have become increasingly problematic. The scientific and technological rationalism of world domination fails because of the exponential environmental risks provoked by the uncontrolled profit-oriented global economy. Social integration in turn is shipwrecked on the rocks of an increasing erosion of the social foundations of communicative action.

6 Intermittent Normativity as Societal Destructuration

After the failure of neoliberal economics in 2007/08, and once more after the global pandemic of 2020/21, fertile ground for the resumption of the welfare state compromise seemed to be given. However, its implementation is faltering, apart from exceptional measures to address economic, environmental and health crises. This is due to the fact that the modalities of interchange between public sector and private economic enterprises established during 40 years of neoliberal policies cannot be changed without consistent investments in resources and human capital to strengthen the state agencies that would be in charge of implementing a new welfare state compromise [35]. Yet, the deeper reason for the missing societal transformation is related to the societal incapacity of normative structuration. This depends primarily on the fact that societal structuration processes have acquired an increasing intermittent character and that the advanced fragmentation of complex societies has prevented the rise of social agencies capable of becoming the bearer of societal transformation. Accordingly, a diagnosis of the missing normative structuration in complex societies is needed, concerning both the institutional aspect of failing social structuration and the fragmentation of social action subjects [5].

In contemporary societies, social structuration increasingly involves a number of relationships with anonymous third parties. The centrifugal fragmentation of social action centres not only presents them with the Sisyphean task of steadily rebuilding, through cultural work, the missing self that society requires of them. It also challenges social structuration with the entropy of the action-coordinating knots required for the legitimation of the legal and political systems. The idling of social structuration processes accumulates a 'distress of culture' that undermines social legitimacy on the temporal axis, so that societal life cycles assume a feverish character, which does not result in normative stabilisation. Instead, structuration processes follow accelerating dynamics whose intermittency substitutes normative social integration.

Complex societies consist increasingly of multiple, alternative and competing perspectives in shaping social structures that fight for predominance. Intermittent legitimation is thus more than ever in demand. Consensus-driven social action spreads as a means to establish provisional asymmetric relationships between social actors, according to which the expectations of the one side are treated as 'valid' by the other side, even without agreement to that effect, yet only within narrow spatiotemporal limits of legitimacy. Social actors regard each other's expectations as 'practically binding' for their (re-)action, whatever the reasons are for accepting those claims. The one side acts on the basis of 'expectations of consensus', and the other side on the basis of 'provisional acceptance of consensus claims'.

A hypertrophic establishment of competing networks of social relationships takes place, which only affects qualitative differentiated fragments of action centres, with very limited scope of stabilisation. Too many different and steadily changing structuration processes compete in the same social space, thus impeding the construction of any consistent normative action framework. The increased number of regulations are not negotiated by social actors, but are imposed based on claims of imputation by decision-making bodies, which are ever more remote from the lifeworld of the persons affected. A dialectics of alienation and passivity conforms to an increasingly short-lived consensus-driven social action legitimation. The vertical asymmetry of social relationships overlaps with the uncontrolled qualitative societal differentiation that characterises contemporary societies. There is a clash of a multiplicity of parallel realisations of social structure that follow different and at times opposing logics, provoking a fierce struggle between different qualitative reified domains of society and leading to continuous phenomena of colonisation.

The creativity, rhythm and tempo of social action replace the declining normative integration of complex societies. What holds the social fabric together is the increased frequency of largely anonymous social exchanges rather than its missing homogeneity. Thereby, complex social reality brings benefits of the highest plasticity and can adapt to the most unexpected and rapid societal changes caused by the restructuration rhythm of crumbling late capitalism. Social heterogeneity can be processed. Yet, societies also increasingly lack a normative foundation that can integrate social relationships into a legitimated framework of solidarity. Qualitative societal differentiation alternates between phases of strong uncontrolled differentiation and rapid regression that implicate a wide overproduction and destruction of human, cultural, societal and natural resources. The 'legality' of social orders and the 'legitimacy' of domination relationships, based on creative performances of 'social validity acknowledgment', only become possible in the form of a tension-fraught intermitting merging of social structure and social action. Taken to its extreme, the conflict between the logic of social action and social structure leads either to a complete alienated reduction of social actors to the logic of social structuration processes (robotisation) or to the blasé rejection of every sociation (privatisation).

Normative intermittency spreads, so that officially acknowledged social orders are increasingly mismatched with the empirical social praxis of legitimation. Failing societal structuration thus jeopardises regulating institutions. The increasing intermittency of social orders makes legally established normativity ever more precarious, so that the rule of law, which controls the recourse to force by its legitimated monopolist, weakens and loses its grip on social reality. In political terms, the normative imbalance of complex societies takes the shape of a loss of democratic control over policy-making decisions. Under the pressure of the temporal urgency of economic, societal or environmental emergencies, the institutional mechanisms of domination established through the monopoly of the legitimate use of force increasingly escape the control of parliaments. Hence, in contemporary societies, the historically evolved domestication of concentrated political power seems to be in serious danger.

7 Failing Social Structuration and Transformative Social Action

Due to the growing social fragmentation in the age of neoliberalism, social and political collective subjects who ensured that the formal recognition of citizenship rights was followed by their substantive implementation, have fundamentally lost their capacity to act. A widely recognised welfare state compromise has been substituted by the parochial representation of egoistic interests motivating single social strata or groups. The deterioration of the implicit social contract in complex societies generates the social, cultural and political consequences that are before everyone's eyes. The winners of the normative societal destructuration are the top 10% highest wealth holders, who in recent decades have achieved almost complete tax avoidance, thereby generating a loss of revenue and sacrifices of the implementation of welfare-states measures.

The increasing rhythm of social restructuration and crisis does not lead to a dialectical overcoming of the dominant development model as the philosophy of history expects. No innovative societal structuration emerges that establishes transformative processes capable of reframing the domination of the economic sector in contemporary society and delimiting its destructive effects on social life and the environment. Social structuration still establishes norms and regulations, which are necessary for society to function, yet with increasingly limited spatiotemporal warranty. Hence, such regulations do not enjoy the legitimation of consistent normative structuration processes, because social actors know that in scenarios of sudden economic emergencies they will adapt to the next compromise of interests. Individual and collective social actors are thus involved in passive and alienated consensus performances, as long as they do not reactivate cultural work that makes them conscious producers of innovative syntheses in social structuration processes.

Yet, the sociological diagnosis is severe concerning the margins of realisation for innovative processes of societal structuration. Fragmented normative action pursued in an accelerated process of hetero-directed societal restructuration with delimited spatiotemporal boundaries of validity, increasingly induces shifting legitimacy. Nevertheless, if we grasp in sociological categories how the dynamics of intermittent normative structuration unfold, we can forecast which forms of 'transformative social action' could possibly lead complex societies to achieve inclusive social integration on the basis of innovative normative structuration processes and conceptions of societal solidarity.

Communicative action, in its classical formulation, cannot unfold in a scenario in which social fragmentation and destructuration rhythms have reached levels of scarcity in spatiotemporal and social resources that do not permit the establishment of an ideal communication situation [36]. The resistance of the lifeworld towards irrational economic imperatives is extremely reduced and does not lead to renewed normative societal structuration. Yet, contrary to what systems theory assumes axiomatically, social system building as an autonomous, continued and self-referential communication flow also fails because of the increased intermittency, multiplicity and reciprocal colonisation characterising the precarious social structuration processes of complex societies [37]. The social transformation that characterises contemporary societies thus generates an unprecedented scenario for both social theory and political practice.

What characterises intermittent normative societies, which are based on asymmetric consensus relationships and do not enjoy consistent social legitimation, is on the negative side: disorientation of social action, colonisation between social spheres and crumbling societal structuration. This *status quo* impedes societies from reaching stable normative arrangements and exposes them to social deconstruction and authoritarian political drifts. Yet, on the positive side, because of their enhanced 'solid liquidity' [38] and intermittent structuration, contemporary societies have reached an unprecedented degree of plasticity, which can be redirected in the sense of a renewed normative social structuration. This latent shaping potential allows the unfolding of societal arrangements that constitute the best breeding ground to realise a just and environmentally compatible society for the twenty-first century. In opposition to the axiomatic assumptions of traditional theories of social transformation, however, a political praxis for the welfare state compromise of the twenty-first century must be constructed in a historic frame that cannot claim its success as a necessary development of societal evolution. More than ever, modernity qualifies as an intermittent destructuring structuration process, so that the horizon of transformative social action finds itself beyond the certainties of the philosophy of history ([5], 246–254). A new social contract for contemporary societies as well as substantial citizenship rights thus constitute a potential, yet not assured, development of the ongoing social change.

References

1. Esping-Andersen, G.: The Three Worlds of Welfare Capitalism. Princeton University Press, Princeton, New Jersey (1990)
2. Fourastié, J.: Les Trente Glorieuses, ou la révolution invisible de 1946 à 1975. Fayard, Paris (1979)
3. Marshall, T.H.: Citizenship and Social Class [1950]. Pluto Press, London (1996)
4. Duménil, G., Levy, D.: Capital Resurgent: Roots of the Neoliberal Revolution. Harvard University Press, Cambridge (2004)
5. Fitzi, G.: Normative Intermittency. A Sociology of Failing Social Structuration. Palgrave, New York (2022)
6. Reeves, R., McIvor, M.: Clement Attlee and the foundations of the British welfare state. Renewal **22** (3–4), 42–59 (2014)
7. Ptak, R.: Vom Ordoliberalismus zur sozialen Marktwirtschaft: Stationen des Neoliberalismus in Deutschland. Leske and Budrich, Opladen (2004)
8. Addison, P.: The Road to 1945: British Politics and the Second World War. J. Cape, London (1975)
9. Beveridge, W.: The Beveridge Report. Interdepartmental Committee on Social Insurance and Allied Services, London (1942)
10. Dutton, D.: British Politics since 1945: The Rise and Fall of Consensus. Blackwell, Oxford (1994)
11. Weber, M.: Max Weber Gesamtausgabe. Mohr Siebeck, Tübingen. Quoted here as MWG followed by the volume number (1984–)
12. Piketty, T.: Capital and Ideology. Belknap Press, Cambridge, MA (2020)
13. Tymoigne, É.: Deregulation, the Financial Crisis, and Policy Implications. Jerome Levy Economics Institute, Annandale-on-Hudson, NY (2009)
14. Kennedy, G.: Adam Smith and the invisible hand: from metaphor to myth. Econ. J. Watch **6**(2), 239–263 (2009)
15. Appel, H., Orenstein, M.A.: Why did neoliberalism triumph and endure in the post-communist world? Comp. Polit. **48**(3), 313–331 (2016)
16. Featherstone, M.: Planet Utopia: Utopia, Dystopia, and Globalisation. Routledge (2017)
17. Piketty, T.: Capital in the Twenty-First Century. Harvard University Press, Cambridge, MA (2014)
18. Boltanski, L., Chiapello, E.: The New Spirit of Capitalism. Verso, London (2018)
19. Emmenegger, P., et al.: The Age of Dualization: The Changing Face of Inequality in Deindustrializing Societies. Oxford University Press, New York (2014)
20. Greve, B.: Welfare, Populism and Welfare Chauvinism. Policy Press, Bristol (2021)

21. Standing, G.: The Precariat. Bloomsbury Academic, London (2011)
22. Watts, B., Fitzpatrick, S.: Welfare Conditionality. Routledge, New York (2018)
23. Böttcher, M.: Der Null-Stunden-Vertrag. Arbeitszeitflexibilisierung nach deutschem und britischem Recht. In: *Beiträge zum Arbeitsrecht.* No. 9. Mohr Siebeck, Tübingen (2020)
24. Hultzsch, F.: Nullstundenverträge. Grenzen arbeitsvertraglicher Flexibilisierungsmöglichkeiten im Hinblick auf Lage und Dauer der Arbeitszeit. In: Schriften zum Sozial- und Arbeitsrecht. Nr. 357. Duncker & Humblot, Berlin (2019)
25. Rubery, J.: Change at Work: feminisation, flexibilisation, fragmentation and financialisation. Empl. Relat. **37**(6), 633–644 (2015)
26. Wolf, H.: Fragmentierte Arbeit im Postfordismus: Übersehene Wechselwirkungen zwischen betrieblicher und überbetrieblicher Rationalisierung in der Automobilindustrie. In: Dunkel, W., Hanekop, H., Mayer-Ahuja, N. (eds.) Blick zurück nach vorn. Sekundäranalysen zum Wandel von Arbeit nach dem Fordismus. International Labour Studies, pp. 135–174. Campus Verlag, Frankfurt/M (2019)
27. Nachtwey, O.: Germany's Hidden Crisis: Social Decline in the Heart of Europe. Verso, London (2019)
28. Simonazzi, A., Ginzburg, A., Nocella, G.: Economic relations between Germany and Southern Europe. Camb. J. Econ. **37**(3), 653–675 (2013)
29. Gallino, L.: La scomparsa dell'Italia industriale. Einaudi, Turin (2003)
30. Pulignano, V.: Atypische Beschäftigung und Fragmentierung des Arbeitsmarktes in Italien – "Karussell der Prekarität"? In: Holst (ed.), pp. 143–164 (2017)
31. Fana, M.: Non è lavoro, è sfruttamento. Laterza, Rome and Bari (2017)
32. Habermas, J.: Legitimationsprobleme im Spätkapitalismus. Suhrkamp, Frankfurt/M. (1973)
33. Marx, K., Engels, F.: Marx-Engels Gesamtausgabe. In: Berlin-Brandenburgische Akademie der Wissenschaften (ed.) Berlin: Dietz, since 1998 Akademie Verlag, quoted here as MEGA followed by the volume number (1975–)
34. Walby, S.: Crisis. Polity Press, Cambridge (2015)
35. Crouch, C.: The Strange Non-Death of Neoliberalism. Polity, Cambridge (2013)
36. Habermas, J.: The Theory of Communicative Action. 2 vols. Beacon, Boston (1984)
37. Luhmann, N.: Soziale Systeme: Grundriss einer allgemeinen Theorie. Suhrkamp, Frankfurt/M. (1984)
38. Fitzi, G.: Modernity as Solid Liquidity. Simmel's Life-Sociology. In: Kemple, T, Pyyhtinen, O. (eds.) The Anthem Companion to Georg Simmel. Anthem Press, London and New York, NY, S.59–80 (2016)

Gregor Fitzi is researcher at the Faculty for Social and Cultural Sciences at the European University Viadrina, Frankfurt (Oder). After his Ph.D. in Sociology at the University of Bielefeld, he was assistant professor at the Institute of Sociology, University of Heidelberg, Interim Full Professor of Sociological Theory at the University of Bielefelsd and Co-Director of the *Centre for Citizenship, Social Pluralism and Religious Diversity*, at University of Potsdam. Among his recent publications are *Normative Intermittency. A Sociology of Failing Social Structuration.* New York: Palgrave, 2022. With Jürgen Mackert and Bryan S. Turner (Eds.), *Populism and the Crisis of Democracy.* 3. Vols. London: Routledge, 2019. With Bryan S. Turner. *Max Weber's Politics as a Profession.* Special Issue of the *Journal of Classical Sociology*, 4/2019.

The Dangerous Adventure of Designing Bubbles

Lidia Gasperoni

Abstract

Bubbles are concretisations of temporary spaces. On the one hand, the bubble constitutes a structural, inflatable element characterised by a specific, malleable, transparent materiality; on the other hand, it represents a conceptual spatial device for expressing a specific feature of contemporary anthropological spaces, a place of exposure, in which the boundaries between outside and inside, between soil and sky, seem to disappear. In this essay, I will first reflect on the ephemeralisation of space, focusing on the definition of space as a network. Second, I will elaborate on this theoretical diagnosis by drawing on work by Vilém Flusser, who explicitly proposes the bubble as a space of contemporary living. Third, through discussing specific examples, I will attempt to develop a critique of bubble spaces as a dimension of contemporary design.

Keywords

Bubble · Network space · Ephemeralisation

L. Gasperoni (✉)
Department of Architectural Theory, Technical University Berlin, Berlin, Germany
e-mail: lidia.gasperoni@tu-berlin.de

1 Introduction

As configurations of temporary spaces, bubbles shape contemporary urban interventions. Publications such as *Bubbletecture*[1] and the recent exhibition *Aerodream*,[2] curated by Frédéric Migayrou and Valentina Moimas at the Centre Pompidou-Metz, deal with the bubble as a central structural element that has been a feature of art, design, engineering structures, and architecture from the twentieth century to the present day. On the one hand, the bubble constitutes a structural, inflatable element characterised by a specific, malleable, transparent materiality; on the other, it represents a conceptual spatial device for expressing a specific feature of contemporary anthropological spaces, a place of exposure, in which the boundaries between outside and inside, between soil and sky, seem to disappear. The origins of the bubble can be identified in that process of ephemeralisation and dematerialisation that emerges from the use of inflatable structures in engineering and architecture and that goes hand in hand with two central phenomena in the development of contemporary industrialised societies: the first is the technical development of inflatable structures at the base of the first airships, which were first built in the mid-eighteenth century and culminated in the use of airdrops and

[1] Reference [15].

[2] Reference [25].

A. Borsari et al. (eds.), *TEMPORARY: Citizenship, Architecture and City*, The City Project 4,
https://doi.org/10.1007/978-3-031-36667-3_3

aircrafts in the Second World War—a war no longer fought solely on the ground but also in the air, in which ephemeralisation and deceiving the enemy were key strategies.

The second phenomenon is the cultural development of modern, globalised societies, in which geographical and material spatial experiences have changed as a result of the increasing ephemeralisation of distances and immediate bodily experiences. This process accelerated with the invention of telecommunications and the Internet. The bubble space therefore emerged from a technical and cultural space of ephemeralisation, the spirit of which can be summarised by Le Corbusier's association of the architectural building with a soap bubble as a space in which "the exterior is the result of an interior"[3] and Buckminster Fuller's conception of doing "more with less",[4] developing experimental geodesic dome structures. In the second half of the twentieth century, bubbles shaped commercial production, finding several uses in industry, agriculture, and social activities. The exhibition *Structures Gonflables* in Paris (1968) and the Expo in Osaka (1970)[5] represent two paradigmatic moments in the use of pneumatic structures and the visions for which they stood.

Bubbles thus became a part of architectural design, as the work of Frei Otto and Frank Lloyd Wright shows. In this period, inflatable structures and bubbles also assumed a crucial role in artistic and architectural experimentation: Haus-Rucker-Co's Oase no. 7 (1972), which features a bubble protruding from an exterior wall of the Museum of Art and Trade in Kassel, is a paradigmatic example. Today, bubbles are often used in temporary architecture, for different—commercial, experimental, and social—purposes, as in the work of Raumlabor and Plastique Fantastique. The bubble is a dimension, at once spatial and social, in which existence is addressed as nomadic and homeless. Temporary spaces, for instance in cities, are reappropriations of places that have been marginalised and excluded from protection, which is still technically and culturally a core dimension of our life and imaginary (in a time of pandemic and war). From this perspective, this essay aims, (i) to present a theoretical framework in which the bubble emerges as a temporary space and, (ii) to reflect critically on its materialisation. I will first focus the analysis of the ephemeralisation of space on the definition of space as a network. Second, I will elaborate on this theoretical diagnosis by drawing on work by Vilém Flusser, who explicitly proposes the bubble as a space for contemporary living. Third, by discussing specific examples, I will develop a critique of bubble spaces as a dimension of contemporary urban design and architecture.

2 Space as a Network: A Philosophical Diagnosis

The definition of space as a network of relations, which has also seen a technical implementation in the digital age, is an analysis, or even a diagnosis, that was introduced in the twentieth century in philosophy at the intersection between cultural and media studies to understand how the concept of space has changed over time. Space can no longer be defined only through the lens of a visually perceivable materiality that shapes the relation between subject and object; instead, it is shaped by new forms of materiality whose structures are invisible and ephemeral. This ephemeralisation of matter is a phenomenon that crosses several dimensions of experience and touches on how we look at biological structures, geographical scales, social attitudes, uses of technical devices, and so on.

In the multimedia exhibition project *Les Immatériaux* at the Centre Pompidou in 1985, Jean-François Lyotard gathered a series of phenomena that have shaped new modes of materiality. He coined the term *immatériaux* to define a type of matter which, in our contemporary age, is no longer material in the traditional sense of

[3] See [22]: "A building is like a soap bubble. This bubble is perfect and harmonious if the breath has been evenly distributed and regulated from the inside. The exterior is the result of an interior".

[4] See [3].

[5] See Aerodream pp. 146–169. Reference [20].

something we can perceive in the presence of the object.[6] In this regard, life today is shaped by coding and decoding systems that "are intangible in a new way. Good old matter itself reaches us in the end as something that has been dissolved into complicated formulas and reassembled. Reality consists of elements organised by structural sets (matrices) in scales of space and time that are no longer human".[7] Space thus becomes a site of complex relations that demands a new sensitivity.[8] The ephemeralisation of space, which can also be defined as a kind of dematerialisation or—according to Lyotard—"*immaterialisation*",[9] is the basis of modernity. According to Michel Foucault, the definition of space as a network of relations marks the third phase in the development of the Western experience of space:

> We are in the epoch of simultaneity: we are in the epoch of juxtaposition, the epoch of the near and far, of the side-by-side, of the dispersed. We are at a moment, I believe, when our experience of the world is less that of a long life developing through time than that of a network that connects points and intersects with its own skein.[10]

Looking back, the first phase of the experience of space was still linked to the possibility of a specific emplacement and hierarchical organisation of places in the Middle Ages. The space of emplacement was accordingly, on the one hand, characterised by the hierarchical organisation of places and the local limitation of the known; on the other, it was a radical experience of leaping into the unknown without a regimented mapping that made the paths familiar in advance. It was a space of local perception and a slow scanning of the world. According to Foucault, the second phase was precisely opened up by Galileo, whose scandal was the constitution of an "infinite, and infinitely open space"[11] in which medieval space dissolved. Emplacement was thus replaced by extension, which defines a place in relation to other points and at the same time differentiates it from the determination of diffuse, subjective sensations and infinite beings. In this period, spatial determination is relativised; it is graspable through relations and not from an absolute, neutral point of determination. The third phase, the phase in which we currently find ourselves, focuses on site, which "is defined by relations of proximity between points or elements; formally, we can describe these relations as series, trees, or grids".[12] Foucault associates the problem of site with, among others, urgent questions in contemporary society such as the storage of data and demographic distribution and relations. Despite its ephemeral turn, Foucault does not view space as completely desacralised and instead describes it as relational: "[...] we live inside a set of relations that delineates sites which are irreducible to one another and absolutely not superimposable on one another".[13]

Space is therefore not an abstract definition or static representation of cultural and social phenomena but their genetic crystallisation, which is a specific configuration of relations. From this perspective, space emerges within a field of practices that shape its cultural and anthropological connotations. This ephemeral and dematerialised space is thus not an abstract idea; it is the result of spatially situated practices that configure the anthropological space that is, to echo Michel de Certeau, graspable only as a "practiced place".[14] Whereas geometric space can be defined through measurable, quantifiable parameters, anthropological space, also defined as existential, is realised in everyday practices and is therefore a genetic space, a culturally given space.

[6] Lyotard, J-F.: After Six Months of Work... (1984), pp. 29–66. 30 Years after Les Immatériaux: Art, Science, and Theory, ed. By Yuk Hui and Andreas Broeckmann.

[7] Lyotard, J-F. with others: Immaterialität und Postmoderne, p. 11. Merve Verlag, Berlin (1985) [author's translation].

[8] See [18].

[9] See [23].

[10] Reference [14].

[11] Ebd., p. 23.

[12] Ebd.

[13] Ebd., pp. 23–24.

[14] Reference [7]. The definition of space as a practised place highlights the impossibility of defining space abstractly and draws on Maurice Merleau-Ponty's distinction between geometric space and anthropological space.

Foucault also highlights spatial practices' power to shape and co-constitute society, and Henri Lefevbre's conception of social space is also worth considering in this context. Manuel Castells likewise analyses the complex relations between space and society, which become visible precisely because space is "not simply a reflection of society, it is its expression. In other words: space is not a photocopy of society, it is society. Spatial forms and processes are formed by the dynamics of the overall social structure. [...] Indeed, *space is crystallized time*".[15] Based on this genealogical consideration of the relation between space and society and on David Harvey's definition of space as a material product in *The Condition of Postmodernity*, Castells introduces the concept of the space of flows, which is of central importance in our network society. These are flows of capital, information, technologies, images, and interactions that are an "expression of processes *dominating* our economic, political, and symbolic life"; in the space of flows, which is "the material organization of time-sharing social practices that work through flows", places disappear "since the positions are defined by the exchanges of flows in the network".[16] Castells analyses how the network is formed by flows and nodes that have taken on a design in concrete space. In doing so, he refers to urban spaces, specifically to certain architectural contexts that the network makes visible in its nodes: global cities are paradigmatic examples of the homogenising tendency of the network insofar as they embody an international architectural design based on common functions and attitudes that are found throughout the world.

The space of the network is a hybrid space in which situated spaces become nodes of the network itself, which is at once a set of trajectories and joining, in-between spaces. The network thus engenders a dynamic of appearance and disappearance. Its complexity, which is also the source of its own invisibility, is made up of nodes, situated and constructed places, and spaces in between, which in turn are not empty spaces or mere filler spaces but spaces that generate the nodes and constitute their trajectories, which are immanently temporal. In this regard, the conception of space as a network is also connected to Paul Virilio's concept of *dromology*, according to which movement, circulation, and communication media shape the spatial and territorial structures of society.[17]

3 The Flusserian Bubble

Flusser shares this diagnosis of the transformation of space as a network of relations. His definition of space as network results from the impossibility of drawing boundaries. This analysis is not merely theoretical, rather a spatial reflection that suggests the bubble as a possible conformation of contemporary space, establishing a reciprocal area of influence between theory and architecture. From a theoretical point of view bubble relates to proximity. This has become a space–time category that leads to the dissolution of boundaries, such that we can be simultaneously virtual-spatial and concrete-temporal in the most diverse of geographical locations. For Flusser, the boundary between the public and the private is thus overcome as a rigid separation. As he writes:

> The only concrete thing is the relations themselves, whereas everything related or in relation (all objects and subjects of relations) are abstractions. The network of relationships is to be seen as an intersection of different relational fields, one of these being the field of intersubjective relationships. The field of dialogical intrahuman relationships is networked with other fields in a way that is almost too complex to understand (for example, with the electromagnetic, with the psychological or the ecological).[18]

Flusser's diagnosis of the transformation of the concept of space is based on the new techniques for its virtualisation. Since space has become calculable and made up of particles, virtual space has surpassed vital space. We experience vital space differently "since we

[15] Reference [5].

[16] Ebd., pp. 442–443.

[17] Reference [28].

[18] Reference [10].

calculate virtual space". This creates a "tricky loop", according to Flusser: "Space is contained by virtual space and virtual space by cosmic space, and vital space is a special case of both cosmic space and virtual space".[19] Both geographical vital space and historical time are virtually transformed; scales lose their analogue status, which today is discussed as the problem of the "clash of scales".[20] Embodied perception changes and is brought to a new materialisation by virtual space: "Virtual space and outer space begin to break into vital space, to partially cover it (overlap), and to overlap each other. This pulls us off the ground (out of the here and now) and forces us to become birdlike".[21] Space, transformed in this way, is accordingly characterised by a new way of perceiving that, by virtue of telematic devices, represents a new approach to phenomena and social dimensions (such as family and people) that can no longer be grasped because—as Flusser points out—"it is no longer possible to draw boundaries". The experience of space is also a temporal one, as Flusser writes in his *Kommunikologie*: "[...] we experience space as the synchronisation of time and time as the diachronisation of space".[22]

Thus, Cartesian space should ultimately be overcome and give way to a liberated flowing space, which is not subject to any boundaries. This virtually borderless space will also be as ephemeral as possible in its concrete design, thus giving the transformed space its current form. It eludes geometry and a ground perspective in order to correspond to a boundless perception that is characterised by "the overlapping, the covering, the interlocking".[23] Space is a "relational field where information is generated, stored, and distributed".[24] Starting from this premise, Flusser's topological perspective on space also changes the image of the spatial organisation of a city that should be designed not merely as a geographical place but as a "fold in the intersubjective relational field".[25] This virtualisation of living space changes the task of all forms of spatial design; it is about "opening spaces within which something is placed that could not be imagined before".[26] Accordingly, even the house as a spatial category loses its architectural and ontological significance:

> The perfect house with a roof, walls, windows, and doors exists only in fairy tales. Material and non-material cables have perforated it, making it like Swiss cheese. The roof sports an antenna, telephone wires pierce the walls, windows are replaced by television, and doors have given way to garage and automobile. The perfect house has become a ruin through whose cracks gust the winds of communication. It has become a shabby patchwork. A new type of architecture is needed.[27]

Flusser's cultural-critical definition of space is not merely theoretical; instead, it is associated with architectural design. Theory shows itself in its critically applied potential; it is not descriptive but makes specific demands on practice. Like urban planners, architects must begin to think of the house from a topological, not merely a geographical, point of view. The house must stop being an "artificial cave" and must instead become a "bending in the field of interpersonal relations" or "a node in the network of interpersonal relations".[28] The creative house must find a new way to articulate the dwelling between the outside and the inside, determined by controlled openings. From this existential-philosophical analysis, it will come as no surprise that for Flusser the house to be designed will have neither roof nor walls, neither windows nor doors:

> Such architecture without roofs or walls, which would be open to all the world – in other words, would consist only of reversible windows and doors – would change our existence. People would no longer be able to duck and hide, and they would have neither foundation nor support. They would have no choice but to extend a hand to others. They would no longer be subjects. There would be no masters from whom they have to hide, but

[19] Reference [13]. See also [12].

[20] See [19].

[21] Flusser, V.: Räume, pp. 277–278 [author's translation].

[22] Reference [11].

[23] Flusser, V.: Räume, p. 284 [author's translation].

[24] Flusser, V.: Designing Cities, p. 174.

[25] Ebd., p. 177.

[26] Flusser, V.: Räume, p. 282 [author's translation].

[27] Flusser, V.: Building Houses, p. 57.

[28] Reference [8].

> neither would they be able to seek their protection. And there would be no nature to threaten them or that they would feel the need to tame. On the other hand, these mutually open houses would spawn an unimaginable abundance of projects. They would be networklike switched projectors for alternative worlds that all humankind would hold in common.[29]

For Flusser, redesigning both city and house means reshaping a post-historical form of life that is not yet clearly recognisable and that finds a possible concretisation precisely in the *bubble*. As he writes: "*Space* can no longer be a flat box that sits on the floor and through which time blows towards the future. Rather, *space* has to be a *bubble* that stretches out into the future".[30] The bubble also stands for overcoming the conception of spatial design as a scaffolding or skeleton "on which life leans in order not to dissolve. Rather, 'space' should be a living skin that absorbs information, stores it, processes it in order to pass it on".[31]

Accordingly, the bubble realises a *generalisation of functions*: The spatial design based on boundary dissolution withdraws spaces from their specific and fixed specialisation and brings them into a network: "It will namely have to open up a multitude of interlocking grey zones shifting in space and time, within which the specialists of the individual spheres will jointly create, store and distribute information".[32] In this context, the concept of a function is also obsolete, and a new form is required for the overlapping and intersection of functions. The architect him- or herself becomes a generalist who understands and reinvents functions within a larger horizon. This spatial and temporal dimension of overlap should become the fourth dimension of in-between spaces, the grey zones defined by Flusser, which are then again investigated by specialists.

The boundary between inside and outside, like that between subject and object, is dissolved, and it is the concept of technique that changes: it is no longer an instrument but rather a method for opening up new possibilities and new worlds in which subjects are no longer in a hierarchical dimension with respect to objects, nature, and instruments. This positivist and progressive vision of technology in Flusser is also linked to a nomadic perception in which the home becomes a technicalised tent and human beings accept their existential exposure, associated with the force of the wind.

Flusser's reflections on the transformation of space as a network also represent a cultural-critical framework for the spatial design of bubbles and living skins, as the four issues (104–111) of Arch+ at the beginning of the 1990s show: the topics covered include "intelligent skins", the redefinition of the city, ephemeral architecture, textile structures, and reflections on the façade. Issue 104 of Arch+, entitled "Das Haus als intelligente Haut" [The House as an Intelligent Skin], deals with the theme of a climate-friendly architecture that aims to build an intelligent house that uses new technologies ecologically. In his essay "Dach- und mauerlose Architektur" [Architecture without Roof and Wall], Flusser underscores that what is needed is a new architecture able to appropriate new technologies, conceiving buildings as intelligent works rather than inanimate ones. For Flusser, phenomenology is the model of thought that underpins the new conception of buildings as "simulated skins": Architects and urbanists must accordingly become "dermatologists".[33] Buildings become places of virtual memory in which information is not only received but also preserved. Paradigmatic in this regard is the design for the European pavilion for Expo'92 in Seville by Spence & Webster/DRU: "In our work, we have tried to take into account these two inter-related aspects of energy-efficient building as well as the beneficial effects of daylight on the health and well-being of the users".[34]

The exterior wall loses its status and becomes an intelligent wall that—like a thin skin—reacts

[29] Flusser, V.: Building Houses, p. 58.

[30] Flusser, V.: Räume, p. 282 [author's translation].

[31] Flusser, V.: Räume, pp. 282–283 [author's translation].

[32] Flusser, V.: Räume, p. 284 [author's translation].

[33] Reference [9].

[34] Reference [27].

Fig. 1 Toyo Ito, Pao II: A Dwelling for Tokyo Nomad Women, SOCKS is a project by Fosco Lucarelli and Mariabruna Fabrizi of MICROCITIES, Architecture Cityscape, Landscape, https://socks-studio.com/2016/02/07/pao-dwellings-for-the-tokyo-nomad-woman-by-toyo-ito-1985-and-1989/, last accessed 23/11/2022

"sensitively and at lightning speed".[35] As issue 107 on textile architectures shows, the concept of the membrane is central to architecture that sees itself as a living skin. Thus Toyo Ito speaks of a "breathing architecture" (see Fig. 1): "Expressiveness and display of force oppress and repel nomadic inhabitants who regard space as a transient abode. In contrast, an ephemeral, simple membrane is not laden with meanings and looks like plastic sheeting on building site scaffolding. It seems to be perceived by people's bodies as stimulating and pleasant".[36] A building's dermatology serves to store information and adapt to the environment. Façades transform into interfaces.[37]

4 The Bubble's Exposures

Flusser is aware that the bubble as a type of network space would constitute a "dangerous adventure", but, as he states, "it would be less dangerous than staying stuck in the ruins we are currently inhabiting"[38]: human beings would risk no longer being subjects, they would be completely exposed, without protection; their identity would no longer be one of isolation but one of radical openness, in which they would become part of a space of projection. Even if he recognises that such an architecture is impossible without addressing the status and design of the grid that protects and limits space, Flusser's gesture is a *con-formist* one in the sense that he conceives of the bubble as an expression that largely conforms to a society that is globalised, telematised and irretrievably ephemeral. This is perhaps also why his reflection had a powerful impact on the architects of the 1990s, who were strongly engaged with transformations in architecture due to digital media in design and communication. Looking into the need to find new ways of inhabiting contemporaneity and thus preventing architecture from becoming a merely nostalgic and anachronistic gesture, but identifying the bubble as the form of such conformism, Flusser's analysis risks excluding the criticisms that the ephemeralisation of networked space brings with it. including also a critical conception of the environmental impact of plastic, the use of which increased in the second half of the twentieth century.[39] In the last part of the essay, these criticisms will be briefly examined.

[35] Reference [2].

[36] Reference [21].

[37] See Arch+ 108 (1991).

[38] Flusser, V.: Building Houses, p. 58.

[39] See [6].

Fig. 2 Raumlabor and Plastique Fantastique, Das Küchenmonument, 2006 (*photo* Marco Canevacci)

Compared to its ontological, symbolic, environmental, and cultural connotations, the bubble thus seems to pose more complex problems, as the anthropological, biological, and genetic reflection by Peter Sloterdijk shows.[40] From this perspective, if it is not to become a mere conformistic gesture of such a society, the bubble should include a critical reflection on a certain notion of transparency that is the fruit of what Dieter Mersch, referring precisely to Flusser, has defined as the "ideology of the digital age" and the "utopia of telematics".[41] As in the case of digital houses and façades, technology makes living itself an instrumental sphere in which data are merely functionalised and exhibited, and in which one voluntarily exposes oneself to constant control.[42] In truth, this transformation of the architectural bubble into a technicalised house, apparently agile but in fact controlled, is also criticised by Flusser himself as being unable to keep the flow of information open.[43] The bubble as a mere technicalised dimension, as in the case of the incremental use of household devices, risks becoming a sort of positivist and technicist tool that is utterly uncritical in its use of data, functions, and information.

Nevertheless, transparency and exposure are two core driving forces at the base of the development of bubbles as situated interventions and in most urban installations since the mid-twentieth century. As an interactive way of dealing with urban boundaries, the bubble has become a space of urban re-appropriation, a means of inhabiting places in the city that constitute border spaces, spaces of rest, without a specific function, often abandoned. Beyond a spectacular and commercial logic, the bubble in this context, as the *Küchenmonument* shows (see Fig. 2), inhabits temporary urban places, becoming a place of sharing and entanglement. In this context, the bubble reactivates urban places: becoming a new node of the urban experience, the bubble reveals in-between spaces.

The bubble is thus a spatial structure of a *performative urbanism* that aims at inclusion, at engaging urban spaces by creating a different sensitivity to the places and the demographic diversity that populates the city, engaging the public dimension of the city. In this sense, the bubble is an active reappropriation that aims to

[40] Reference [26].

[41] Reference [24]. See in the same book also the essay by Docherty, T.: The Privatization of Human Interests or, How Transparency Breeds Conformity, pp. 283–303.

[42] See [17].

[43] Reference [9].

Fig. 3 Plastique Fantastique, #StayOut. A playful experiment between isolation and communication, Berlin 2021, *photo* Marco Canevacci

overcome the rigid logic between inside and outside, which also includes how human beings relate to natural elements—as in some works by Plastique Fantastique, where bubbles become experiential spaces within forests or rivers.[44]

This anthropological space is an experiential space determined through practices that constitute a kind of "labour" of the everyday operations described by de Certeau, which constantly transform spaces into places, and vice versa.[45] Bubbles as experimental situated practices thus have the power to transform generic space and to renegotiate its conventional uses, for example in experiencing urban space. In the temporary logic, the bubble is a playful and experiential dimension in which architecture constitutes an environmental and urban mediation by creating a kind of intrahuman relationality and by offering effective temporary solutions, e.g. during pandemics, migration crises, and as lighting devices, where the bubble is used as a cross-boundary space. As a temporary, ephemeral and malleable shaping, beyond a spectacularising logic, the bubble has the critical potential to confront us with the contemporary condition of a liquid society that, according to Bauman, "cannot keep its shape or stay on course for long"[46] and, as a reaction to this dissolution, constitutes heterotopias, borders (e.g. the many urban configurations that take the form of trenches, protective spaces, gated communities, constant surveillance devices).

Without this reflection on heterotopic boundaries, the bubble as a node in a network of relations risks falling prey to a manipulative rhetoric of transparency and fluidity. The bubble in our times still seems to confront its opposite: the intimate, protected, embedded space of the bunker. Consider, for instance, contemporary phenomena such as the reconfiguration of the home during the coronavirus pandemic[47] and the image of Kiev's underground in March 2022 during the bombings: the home during a pandemic is a bunker offering protection from the virus but also includes the bubble of the global network (which also provides temporary protection) (see Fig. 3).[48] Contemporary nomads—whose migration is generated by

[44] See [4].

[45] De Certeau, M.: The Practice of Everyday Life, p. 118. See also the analysis of tour and map, pp. 120–121.

[46] Reference [1].

[47] See Reference [16].

[48] See also Designboom https://www.designboom.com/design/plastique-fantastique-mobile-pps-personal-protective-space-doctors-covid19-04-20-2020/, last accessed 23/06/2022.

broken economic, political and social global bubbles—are exposed, devoid of protection, living in bubble tents. During the bombings, the bubble of a telematic, digitally connected, underground, invisible space is transformed into a bunker space of survival.

Thus space as a network of relations should not become a bubble that evades reflection on the boundaries and spatial configurations between its nodes. The definition of space as a network cannot be separated from its cultural-critical diagnosis of Foucault's heterotopic spaces and panoptical structures, as avoiding that space as a bundle of relations generates a flat social ontology without any analysis of the structures that generate them. If it is to avoid becoming a social bubble, the bubble that seeks to reactivate urban spaces as places of exchange and difference is entrusted with the task of constituting a visible and perceptible space of ephemeralness which is also a space of fragility. As Sloterdijk reminds us the bubble seems to be destined to blow up and become part of a plural world.[49] The bubble space cannot exist in the absence of reflection on the boundaries that it seems to want to make transparent.[50] The bubble, even in the guise of a minimal architectural intervention, is a symptom and device of the border, a dangerous adventure, a reflection on real places of coexistence in the network space.

References

1. Bauman, Z.: Liquid Life, p. 1. Polity Press, Cambridge (2005)
2. Breukelman, A.: Architektur der variablen Häute. Arch+ 104, 55 (1990)
3. Buckminster Fuller, R.: Intuition, p. 68. Doubleday & Company Inc., Garden City, New York (1972)
4. Canevacci, M.: Interview 11/Plastique Fantastique. In: St Hill, C. (ed.) This is Temporary: How Transient Projects are Redefining Architecture, pp. 167–175. Riba Publishing, Newcastle upon Tyne (2016)
5. Castells, M.: The Rise of the Network Society, p. 441. Wiley-Blackwell, Oxford (2010)
6. Davis, H.: Plastic: accumulation without metabolism. In: Hannah, D., Krajewski, S. (eds.) Placing the Golden Spike. Landscapes of the Anthropocene, pp. 66–73. Publication Studio. Portland (2015)
7. De Certeau, M.: The Practice of Everyday Life, p. 117. University of California Press, Berkeley, Los Angeles, London (1988)
8. Flusser, V.: Building houses. In: Flusser, V. (ed.) The Freedom of the Migrant, p. 58. University of Illinois Press, Urbana and Chicago (2003)
9. Flusser, V.: Dach- und mauerlose Architektur. Arch+ 104, 68 (1990)
10. Flusser, V.: Designing cities. In: Ströhl, A. (ed.) Writings. Vilém Flusser, p. 174. University of Minnesota Press, Minneapolis and London (2002)
11. Flusser, V.: Kommunikologie. Umbruch der menschlichen Beziehungen? In: *Schriften*, vol. 4, p. 218. Bollmann Verlag, Mannheim (1996) [author's translation]
12. Flusser, V.: On Three Spaces. Artforum. Essays, pp. 205–216. Metaflux Publishing (2017)
13. Flusser, V.: Räume (1991). In: Dünne, J., Günzel, S. (eds.) Raumtheorie. Grundlagentexte aus Philosophie und Kulturwissenschaften, p. 277. Suhrkamp, Frankfurt a.M. (2006)[author's translation]
14. Foucault, M.: Of other Spaces. Diacritics **16**(1), 22 (1986)
15. Francis, S.: Bubbletecture. Phaidon Press Limited, London (2019)
16. Gasperoni, L.: The bunker-bubble. Reflecting the unreal time. In: Armstrong, R., Breddels, L., Gantois, G., Ooms, T., De Smet, A. (eds.) Design in Unreal Times, pp. 34–40. KU Leuven, Leuven (2020)
17. Han, B.-C.: The Transparency Society. Stanford University Press, Stanford (2015)
18. Hill, J.: Immaterial Architecture. Routledge, London, New York (2006)
19. Horn, E., Bergthaller, H.: The Anthropocene. Key Issues for the Humanities, p. 102. Routledge, London and New York (2020)
20. César, I.L.: World Expos. A History of Structures, pp. 378–435. Architect Publications S.L., Barcelona (2017)
21. Ito, T.: Atmende Architektur, Arch+ 107, 55 (1991)
22. Corbusier, L.: Towards a New Architecture, p. 181. Dover Publications, New York (1986)
23. Lyotard, J.-F.: "Les Immatériaux". Epreuves d'écriture. Editions du Centre Georges Pompidou, Paris (1985)
24. Mersch, D.: Obfuscated transparency. In: Alloa, E., Thomä, D. (eds.) Transparency, Society and Subjectivity. Critical Perspectives p. 276. Palgrave Macmillan, Cham (2018)

[49] Reference [26].

[50] This conception could also be compared with an environmental definition of the space of intersection between bubbles, between human and non-human environments, that constitutes the "life space"—according, for instance, to Jakob von Uexküll.See Theoretical Biology, Forgotten Books, London (2018).

25. Migayrou, F., Moimas, V.: Aerodream. Architecture, Art, Design et structures gonflables. Éditions HYX, Orléans (2021)
26. Sloterdijk, P.: Bubbles. Spheres, Vol. I: Microspherology. Semiotext(e), Los Angeles (2011)
27. Spence, R.: Gebäude mit variabler Haut. Arch+ 104, 37 (1990)
28. Virilio, P.: Speed and Politics: An Essay on Dromology. Semiotext(e), New York (1977)

Lidia Gasperoni is a research and teaching associate at the Department of Architectural Theory of the Institute of Architecture at the Technical University Berlin. She studied philosophy in Rome, Freiburg, Breisgau, and Berlin and obtained her PhD from the TU Berlin. She teaches architectural theory and philosophy with a focus on media philosophy, Anthropocene theories, and aesthetics at the TU Berlin, and previously at the UdK Berlin and the University of Kassel. Some of her publications: *Versinnlichung* (De Gruyter 2016), *Media Agency* (with Christophe Barlieb, transcript 2020), *Artefakte des Entwerfens* (with Anna Hougaard et al., TU Verlag 2020), *Construction and Design Manual: Experimental Diagrams* (DOM publishers 2022), and *Epistemic Artefacts* (with Matthias Ballestrem AADR 2023).

Whose Shadow? On Camps and Counter-Camps

Federico Rahola

In girum imus nocte et consumimur igni
Tito Lucrezio Caro/Guy Debord

Abstract

Confronting with the current proliferation of administrative detention facilities for displaced and illegalized persons, the paper traces the different manifestations of this specific border apparatus back to a more general and abstract "camp form"—one whose origins date back to the colonial realm, finding in the colonial subject the first internable figure. To such institutional form, it opposes the relentless production of informal, occupied, and often clandestine encampments, dwelled by people on the move and scattered along as many hidden and illegalized routes. By rereading current makeshift camps and hidden routes though the historical lens of the US pre-civil war Underground Railroad, the article suggests to conceive of them in terms of as many counter-spaces of a possible "Underground Europe" whose material and unauthorized existence, often supported by criminalized solidarity networks, mirrors and reverses that one of a "Borderland" or "fortress Europe". By the same token, it suggests to conceive of these temporary, precarious and informal zones in terms of as many "reverse shot" of the institutional "camp-form", defining them as counter-camps and focusing on the political, spatial and temporal relation between these two opposed polarities.

Keywords

Administrative detention · Camp-form · Border politics · Informality · Dwelling policies · Postcolonial theory · Underground railroad

1 The Camp as a Form

According to the data provided by the Global Detention Project (www.globaldetentionproject.org) there are currently more than 1360 officially registered detention centres for displaced people scattered all around the world, 420 of which are in Europe. Obviously, each of them may strongly differ in size, conditions, population, and purpose. Yet, the word "camp" seems to be

F. Rahola (✉)
Dipartimento Di Scienze Della Formazione, Università Di Genova, Genova, Italy
e-mail: federico.rahola@unige.it

A. Borsari et al. (eds.), *TEMPORARY: Citizenship, Architecture and City*, The City Project 4,
https://doi.org/10.1007/978-3-031-36667-3_4

comprehensive enough to describe the specific political significance of these spaces of confinement. And it seems to work as a kind of paradigmatic figure, a signifier that, in all its different declinations, finds a broader political matrix in imposing a form of detention that has nothing to do with individual acts or direct responsibilities, but rather affects or defines the very political or "ontological" condition of the detained persons. This condition may in turn be conceived of as both a lack of something and as a peculiar form of political excess.

The re-emerging of administrative detention has increasingly marked the last three decades [2, 3], after a prolonged period of apparent oblivion due to the caesura of concentration (and extermination) camps that definitely marked the short XX century. Yet, if concentration camps and the detention of civilians for security reasons found an antecedent in the colonial realm [7], camps never disappeared in post-war and post-colonial times, and they rather beat, as a dark soundtrack, the rhythm of the pretendedly smooth transition towards a new enlarged or globalised geography.

In a book written almost twenty years ago, I suggested the notion of "camp-form" to define the political significance of a specific confinement apparatus whose genealogy dates back to the dichotomous geography of imperial colonies (finding in the colonial subject the first internable and deportable figure), and whose global spread invites us to problematise the overcoming of that overdetermined geography in the postcolonial present [32]. In the wake of a rather categoric, probably Arendtian, idea of citizenship as a sort of *finis terrae*, and of Giorgio Agamben's notion of bare life [1], that book proposed an interpretation of camps as quite direct manifestations of a form of sovereign power acting in terms of exception upon various subjects who, for various reasons, are considered "out of place", exceeding any form of political belonging. Yet it did miss something. Although insisting on the exceeding (rather than static and exceptional) character investing both camps and the subjects destined to find in them their sole possible territory, it ended up underestimating the direct productive or "positive" dimension of camps, as places where a power and a difference are basically produced rather than pre-assumed or presupposed. It was therefore a matter of focusing on such a productive dimension, one which could be proper to a *dispositif*, and of defining the condition of internable, deportable and exploitable subjects whose life and presence, rather than being denied or excluded, are differentially included or harnessed, sometimes trapped, within a given political and economic spatiality, thus further decomposing its alleged homogeneity [12, 13, 33]. In other terms, it was a matter of shifting from a static paradigm of sovereignty and exception into a more productive, dynamic, assembled and intertwined governmental dimension [20]. But that too was not enough: it was still a partial history, since a form could always be a reaction to something that exceeds it.

To see camps as governmental apparatuses, besides countering their interpretation as direct instances of a space of exception, and the ultimate and paralysed condition of bare life it suggests, could also lead to conceive of them as tools to territorialise and direct any form of "bad" mobility, by forcing and guiding it. Accordingly, if camps' very existence directly signals and indeed reifies a form of unauthorised and "bad circulation", it does so in order to react to it, imposing a pace on it, providing it with a regular, official and institutional place, and enclosing it within forced routes and corridors. From this standpoint, camps are apparatuses that manage illegalised forms of mobility by channelling, shifting and also exploiting them, by governing mobility through mobility [36]. Yet, these forced routes, overdetermined by a kind of logistical rationale made up of hotspots, corridors, special zones, and a myriad of containers piled up inside detention/reception centres, are in turn triggered, relentlessly challenged and subverted by that very "bad circulation", through different underground routes and different informal and illegal camps. In the next pages I will mainly focus on the spatial, temporal and political relations/oppositions between these two polarities.

2 The Impossible Map of Europe

The European map provided by the Global Detention Project gives back an incredibly dense constellation of camps, partially connected, a veritable *archipel de la detention* [9]. And it also suggests that camps' scattered localisation, as direct but not immediate border apparatuses, directly (re)produces a striated geography of Europe, which in turn is surrounded by increasingly harsher external borders.

The outcome of such a double-sided process could be represented as the constant reassembling of a specific "border-scape" (Mezzadra, 2004; [38]), one made up of walls, border islands, fences and barbed wire, overcrowded reception, identification and detention centres, forced corridors, and centralised biometric archives, but also of makeshift camps, illegalised routes, and deferred northbound passages. Thus, the impression we get is above all of a battlefield, a kind of "drifting" within and around European borders, which is reflected in the metonymy of a sequence of names and places: Idomeni, Lesvos, Chios, Lampedusa, Ceuta and Melilla, but also Calais, Ventimiglia, Briançon, the Brenner; or Bihac, Lipa, Velika Kladusa, Sid, Narewka, and so on. The optical effect is thus one of an impossible map, an incoherent sum of scattered dots that can coagulate into an overall discourse of closure, but that still remains contended, fragmented, and marked by a series of frictions.

If we look at it from the outside, from its external borders, such a geography acquires a possible definition by suggesting and even reinforcing the (*vexata*) impression of a "fortress", of a space that is impermeable as much as it is incrementally selective, sealed at its margins by mobile, peremptory and violent land and water borders.[1] To tragically summarise this shielded dimension, which forces people on increasingly risky and congested journeys and crossings, the European space has turned out to be a deadly one, surrounded by an enormous (but somehow measurable and measured)[2] number of corpses, of shipwrecks with or without spectators and "left-to-die boats",[3] of bodies caught in the stifling mesh of its borders. These are precisely "border deaths", which are far from accidental, and are as widespread and more frequent as they are structured and "state sanctioned". Such a backdrop gives us the sense of a more general "necropolitical" space, which Achille Mbembe invites us to consider as complementary to the biopolitical one of governmentality [26]. It is possible to find here something akin to a sovereign exercise or power, though in the indirect sense of an act which produces a fundamental indistinction between life and death and, therefore, of a decision whose effects are distant, located in a grey area, as if to dissolve any partitive and geometric opposition between letting and making die. This is because the borders of (post)colonial Europe, be they external or internal, have always killed like this; and it is this essentially racist dimension, as "state-sanctioned or extra-legal production and exploitation of group-differentiated vulnerability to premature death" ([18]: 28), which continues to emerge from the impossible map of European borders.

Yet, there is another complementary image that must be over-impressed to the one of a "fortress", and another geography of Europe that has been produced in the last two decades. Both of them may appear less spectacular or peremptory if compared with the first and, nonetheless, they turn out to be even more incisive and decisive in their ability to shape and redesign the

[1] This impression is also the product of the extension of readmission agreements with third countries (Turkey, Tunisia, Niger, Egypt, Nigeria, Morocco, and so on) which shift European borders, making them work ahead of time and from a distance; of a maritime control whose primary objective is not to rescue lives but rather to force them to be repatriated (to Tunisia, to Morocco, to Libyan and Turkish concentration camps or directly to their countries of origin); of the adoption of a "hotspot approach", with identification and selection centres and camps located directly on the external borders, which imposes a logistic rationality of preventative, differential, and hierarchical filtering on those who arrive or land. See Garelli and Tazzioli [17].

[2] http://www.unitedagainstracism.org/wp-content/uploads/2017/06/ListofDeathsActual.pdf.

[3] Heller et al. (2012). Report on the "Left-To-Die Boat", http://www.forensic-architecture.org/wp-content/uploads/2012/06/FO-report.pdf.

physiognomy of Europe and to selectively capture, differentially and hierarchically, those who attempt to cross it. It is an image or a geography which, starting from the proliferation of borders between and within single member states, conveys and translates the profoundly uneven nature of these territories, their being torn apart and splintered into a series of islands, enclaves, buffer zones, corridors, and camps which coincide with the irregular and striated space experienced, crossed, and inhabited by an illegalised population in transit. To define the political quality of this space, Etienne Balibar has resorted to the term "borderland" [8], alluding to an extended and potentially deferred border zone which is fragmented and racialised, marked, as it is, by a proliferation of devices and border signals by virtue of which any idea of fortress ends up looking inadequate—if anything, because it shatters.

It is in the reciprocal interference and tension between the external projection of the image of a fortress and the actual internal production of a borderland, that the impossible map of "Europe" must be situated. Such a selective and striated space, populated by just as many hierarchised presences (which Sandro Mezzadra and Brett Neilson define as a "multiplication of labor", mostly built on "race management"), can be read as the generalised outcome or the reflection of a specific "sovereign machine of governmentality" [27]. The fact is that we must preliminarily imagine this machine and the proliferation of borders defining it also as reactions to something that "anticipates" or, in any case, determines them, establishing a specific relation. To use an image, a plateau among the thousand defined by Deleuze and Guattari, we could imagine an abstract, heterogenous "capture apparatus" which responds and shapes itself according to a specific and equally heterogenous "nomadic machine" and, thus, investigate the peculiar assemblage, the isomorphic twine that the two plateaus end up establishing, reshaping one another [14]. Accordingly, the image of the fortress ends up taking on the features of a spectral hologram or the more material and violent ones of a criminal deterrent against a relentless movement towards Europe, while the materiality of a borderland those of a reaction against underground routes and informal encampments which keep being redefined and modified by challenging and contesting it.

3 Being Absconded

In a recent book, written together with Luca Queirolo Palmas, we kept this impossible map before our eyes while seeking to reactivate a story from the past and the promises it contained. Our intention was to project the partially unredeemed history of the underground railroad or UGRR (the abolitionist infrastructure that supported runaway slaves in their escape from the chains of the plantations in the pre-Civil War United States—[10, 15] to the European borderland of the present, in search of a possible "Underground Europe" [31]. In this attempt we were not alone nor the first. Other scholars and activists have resorted to the UGRR story as a possible way to name, explain and redeem the European present [35]. Our way of re-reading that glorious as well as betrayed history may slightly differ from that direction (since the past, too, still needs to be redeemed in the present), but it has allowed us to encounter a series of words (routes, abolition, coalition, but also slavery, and chains) that to a certain extent need to be problematically updated and translated.

Tracing possible though complicated "intimacies" and resonances between the past and the present, the here and there, we turned around a direct and almost ontological question: "What does it feel like to be absconded, fugitives, and hunted"? We suggested that we would have to ask Frederick Douglass, Harriet Tubman, and those runaway slaves who, like them, led or embarked on "a flight to north" in the years of the US Fugitive Act, just to discover how their answer, often reported in the first person, would not be very different from the one we might get today from the migrants, refugees, and undocumented persons caught in the European borderland and classed, by the prose of European "counter-migration" governance, as subjects of unauthorised "secondary" or internal movements. If the former described nocturnal itineraries and daytime breaks in swamps or "safe houses", journeys driven by the polar star of a "perceived freedom" [35], and improvised encampments and maroon communities built and inhabited on the run, the latter tell us about routes that are just as hazardous, driven by a similar aspiration—that is, punctuated by a similar alternation between light and shadows, a similar regime of (in)visibility, and often organised around temporary places that are more or less improvised.

This is because, in actual fact, as far as it is filled by detention camps, the European borderland is strewn with temporary camps, for the most part "clandestine", autonomous, and semi-hidden, built overnight and inhabited by people in transit who, at best, may count on the support of a series of other subjects and groups acting in solidarity that, in turn, are often hindered and persecuted by national and European authorities. In the majority of cases, these are extremely precarious places that sometimes reach exorbitant sizes, mostly growing around, next to, or in the shadow of overcrowded official and emergency centres, like in the case of Moria and its double, Pikpa, on the island of Lesvos. However, these "spaces in the shadows", evoking in a way the Maroons' encampments of the XVIII and XIX century Americas, often tend to proliferate in regions located at the margins, surrounding the external borders, along specific routes (in the far west, like in the case of the Gorougou forest, just behind Melilla, the Spanish enclave in Morocco; or in the Balkans, in Bihac, Sid, Lipa and Velika Kladusa; or even further east or south, along the Greek border with Albania, Bulgaria, and Turkey, or the Polish one with Belarus) which point to just as many new passages, therefore triggering further encroachments by the European states' capture apparatus.[4] In the majority of cases, these "absconded encampments for absconded people"[5] seem to be generated through a process of "meiosis", of doubling and proliferation, going on to configure the correlative, which is often a harsher, more precarious and temporary of the "camp-form" (as a sort of offscreen or side effect). However, as "off-screen", they can settle more or less fleetingly also in the very centre, at the heart of Europe, emerging in the public space of major continental cities (such as Paris, Brussels, Berlin, Athens, Milan, or Rome), in parks and squares, in the vicinity of train stations and transit areas or in the rusty shadows of abandoned areas. In any case, they are more frequently built and assembled close to an internal or international border (for example, in Calais, Ventimiglia, Briançon, Bolzano, Trieste, Gorizia, and so on) as crossroads or waiting rooms subject to the turnover of a heterogenous (mostly male), temporary, and ever-changing population of subjects in transit, many of whom are bound to be violently removed and dispersed once they have been deprived of any further opportunity to move, with the few means available to them (mobile phones, tents, blankets, even shoes) confiscated or destroyed.

The different size, temporality, and localisation of these "eruptions" makes it hard to keep them within an overarching discourse: no categories then, no shape or form in a broad sense. Indeed, it is possible to identify a series of fractures and substantial differences within this particular "informal form". In some cases, as already mentioned, the impression is that they work as emergency solutions, improvised and precarious,

---

[4] During 2019, in order to hinder these "secondary" passages (which, however, exceeded the territories of the European Union), Frontex, the European agency designated to control external borders, started to operate directly in Serbia, Bosnia, Kosovo, and Albania, as well as in Croatia and Slovenia. By violating the limits of its mandate and legitimate radius of action, it directly intercepted, pushed back, or repatriated, with the support of national police forces and armies, thousands of migrants and asylum seekers in transit through Greece, Turkey or Bulgaria who were taking refuge in the squares, stations or markets of Sarajevo and Belgrade, in towns and villages on the Croatian border or in places deeper into Kosovo and Albania, along the rivulets generated by the reconfiguration of just as many possible itineraries [29].

[5] In the very etymology of the word absconded resonates the Latin term *absconditus* that for Ernst Bloch may suggest a further, original and inedited dimension. See Bloch [11].

acting as a mimetic supplement, in the shadows, or as an explicit exit strategy with respect to official reception and containment facilities. In other cases, however (like in Calais, Ventimiglia, or Bihac), their very existence seems to trigger and produce (and almost incite) that of institutional centres, making of the latter the shadow by turning them into the reaction through which to reimpose (mostly through a container, the new logistical "modulor") an otherwise intolerable use of public space and an otherwise ungovernable movement made up of illegalised routes and ways of inhabiting them.

In fact, despite being hidden and invisibilised, these camps are actually placed under the everyday spotlights of a form of control which is obsessive as much as it is intermittent and normally external, from a distance. Scrutinised like fires seen from afar, debated and contested by the "public opinion" (still from a safe distance), they are systematically stigmatised as a threat to security, as scandalous instances of illegality, abuse, danger, and insecurity. They are thus labelled as unhealthy, indecent, "indecorous", uncivilised, and primitive, and therefore alternatively defined as jungles, ghettos, forests, and habitats of (un-)urban outcasts.[6] A peremptory process of racialisation affects these places as much as the subjects that inhabit them, as if to weld them together, suggesting the existence of an appropriate space, the "right place" for a detained and invisibilised race inside the containers of an official camp (or chained, as transit labour, in a plantation nearby). Thus, it is not surprising that these informal spaces are constantly under siege and systematically dismantled, that their inhabitants are evicted and relocated to institutional reception and detention centres even hundreds miles away through spectacular police and ethnic cleansing operations which are often hyper-mediated, justified in the name of public health and safety. Frequently, in particular when they are located on the margins of Europe, these camps are simply removed once the media attention has fizzled out, to then be rebuilt a few days later and a few metres away, in further shadow zones.

Yet, if we take a closer look at them, these makeshift camps may turn out to be specific crossroads, that is, temporary crossing and meeting zones where different biographies, itineraries, and languages concentrate and sometimes mix, sharing, in an overall precarity, spaces, resources, experiences, and knowledge based on more or less known or secret word of mouth and maps. As much as they are segregated and often crossed by internal tensions reflecting "ethnic" or national borders and conflicts, they can produce a specific economy and currency of their own, one made of barters and various services (legal or not), and even a specific language, a vernacular that tactically reflects, like a form of mimicry, that of European governance and is filled with words like "asylum", "fingerprints", "Dublin", "humanitarian or subsidiary protection", or acronyms like CAS, CARA, SPRAR, CAO, and so on. On these ambivalent terms (where material and symbolic violence are diluted in a tactical or mimetic reappropriation), they end up forging a specific relationship with the outside world by filtering and retranslating it. Such a relationship is then deepened and amplified by the fact that these are often specific contact zones, in so far as their material survival, if not their actual existence, is mostly the result of cooperation between migrants, refugees, undocumented persons, and members of NGOs or groups of activists, militants, and volunteers who seem to "reactivate" the abolitionist networks of the UGRR that had arisen to support the revolts/escapes of runaway slaves by challenging various intimidations and a more general criminalisation [31, 35].

These forms of cooperation and of more or less ephemeral coalitions can be characterised by different intensities and found at different latitudes—from transient distribution of food or medical assistance in places where people are exclusively or hopefully just "passing by", down

[6] This also applies when these encampments are not directly linked to border apparatuses, as in the case of the "ghettos" nearby labour intensive (and almost slave) agricultural districts, like in the south of Italy and Spain, whose inhabitants are often also guests of official reception/detention centres [5].

to more organised and stable situations. This is true of the Roja Valley, a sort of "autonomous republic of hospitality" which was created in an inlet on the border between Italy and France. Another example which is in many ways exceptional but by no means unique, is the City Plaza, a former four-star hotel located right in the centre of Athens. After closing down in 2014 due to the effects of the crisis and austerity, it was reoccupied two years later by a "coalition" of activists, migrants, and asylum seekers who turned it into a peculiar "station", where different subjects of secondary movements who had mostly passed through the islands of the Aegean Sea or through the northern border with Turkey (but also some who had relocated to Greece from other European countries) were simply able to "stay", to reorganise their own journeys and their own routes while even trying to reinvent the surrounding public space [25].

4 Counter-Camps (-Spaces, -Conducts)

There would be much more to say about these informal and occupied spaces, just as many ostensible stopovers or stations of a possible Underground Europe, more data to provide and stories to tell. Besides, many of them are already relatively over-examined places—from an ethnographic [4, 30], journalistic, and even architectural perspective. In this regard, to conceive of them in terms of as many potential and possible cities or, worse, as forms of a spontaneous and instant urbanism (as has been done for the Calais "jungle"—[37]) does not seem the best way to comprehend their sense and make sense of their harsh, violently precarious and above all temporary conditions. Yet, although tough, they are actual, tenacious, and even inventive expressions of a specific practice of inhabiting and crossing a territory, as well as material actualisations of a specific intensity, that one supporting a series of routes pointed toward an abstract yet possible space generically named "Europe". In order to grasp this drive, it might be useful to look at them through a specific cartographic gaze, from above, in the attempt to situate them geographically and politically by understanding the (in)visible thread that holds them together. In other words, it is a matter of asking how the assemblage of itineraries, directions and journeys that coalesce in those encampments clashes with a specific border regime.

We should begin with a premise, perhaps a correction, by suggesting, for instance, how even from afar, from above, these spaces do not limit themselves to pointing to an "offscreen", the informal correlative of a given form, of a camp-form, functioning rather as its peculiar "reverse shot", as *counter-camps*. Secondly, we should suggest that this temporary and precarious dimension, which is nonetheless oppositional and contrary, ends up becoming loaded with a more general, geographical and political meaning, something akin to a breach, crossed and inhabited, which is the expression of other directions within and against the European borderland. This is because, in the impossible map they outline, these "temporary autonomous zones" and provisional ways of occupying and inhabiting transit spaces reflect just as many drifts which, most of the time, are continuously improvised and remodelled depending on the capture apparatus that looms over them. However, taken altogether they may convey the sense of a "subterranean geography" that cuts through and rearticulates the European borderland.

Thus, what emerges is the relationship between all these dots or "spaces of representation" and the objective and violent "representation of space" synthesised in the idea of a borderland (strewn with reception/detention/identification camps and with the borders hypostatised by these material devices). It is a sort of "dance" which is macabre as well as often lethal (dotted with frozen bodies "off the beaten tracks" of the Balkan route, caught in the barbed wire and in the nets which, like a "litany", surround the border zones, wounded by the national police dogs, mangled on the rail tracks of a Eurotunnel, and asphyxiated in a container), and which nonetheless is continuously "relaunched" at that very high price. The relationship between the routes dug by the absconded and illegalised actors of unauthorised movements and the assemblage of border devices triggered by them, the modernised "chains" through which states (and something else that we cannot but call capital) attempt to govern, contain, or exploit their presence, generate a sort of loop—a set of "situations" which shape one another without interruptions, without a final word (just the first one, "movement"), turning into the unstable, spectral, and continuously challenged image of a borderland. This has the bitter and stifling taste of a trap, but it can also be a crossroads.

There would be much more to add, also with regard to this last image. For instance, we should explore the specific "logistical reason", summed up in the shape of a container (and dotted with corridors, hubs, digitalised platforms and archives, hotspots, and special zones governed by specific synchronised operations that impose orderly flows and selected geographies—[28]), which invests the mobile assemblage of borders and is reflected in the prose of European and global governance of migrations. And we could also examine in more depth the relationship that it establishes with a possible "counter-logistical" reason animating makeshift camps and those unauthorised movements and spatial practices defined as "secondary", finding in the latter the former's possible triggering motive. This could perhaps be a way of reading all these underground locations in terms of another detour or "drifting".

In girum imus nocte et consumimur igni (roughly translated as "We go round and round at night and are consumed by fire") is a famous palindromic line that dates back to an alleged ancient Roman anonymous poet (maybe Lucretius), as well as the title of a 1978 Guy Debord movie, an attempt to depict some situationist excursions within and against the pre-ordered and organised space of the modern metropolis. It suggests a peculiar, often violent, feeling of "being lost" and absconded which goes from being a condemnation to turning into the possibility of another space and another gaze. And it could also be the condition and position for a specific Maroon perspective and thus for a decentred and marginal point of view, a "situated epistemology" [21] which can be used as a starting point to "destabilise any vision that fixes Europe within racialised and normative imaginaries of nation-states and national identities" ([6]: 437). Through similar lenses, in *Underground Europe* we proposed a more intentional reading of this absconded position and condition, in the filigree of a "set of junctions, networks, and intersections" all located along a specific route, and guided by a specific abolitionist drive. After all, it is precisely this deferred route which may represent the (in)visible thread that connects these dots into an impossible map, the chaotic assemblage of counter-camps within and against European borderland. And it is on these premises that we are taken back in time to the experience of runaway slaves, to their rebellions, voices and alliances which resonate and echo into the present [23]. More precisely, we are taken back to the ability to organise and inhabit the escape by creating a counterspace.

As Michel Foucault once suggested, the exercise of modern power is also (if not primarily) a matter of conducting, that is, of governing by separating bad and good circulation [16]. European borderland, with its *archipel de la détention* based upon a specific "camp-form", immediately responds to this rationale by filtering, selecting, containing and more generally governing mobility through (forced) mobility [36]. Again in Foucauldian terms, a possible Underground Europe, built up of connections, illegalised routes and makeshift counter-camps, might be conceived of as the specific counterspace produced by forms of heterogeneous and collective "counter-conduct" [16], that is, by the effort of "conducting themselves in a different way" [34], even coalescing into as many "utopian margins" built up by all those who are "unavailable for servitude" [11, 19]. In this perspective, counter-camps, even without any intentional will to come out from the shadow as possible dystopic cities, yet making of institutional camps a shadow, might represent concrete instances of what Stefano Harney and Fred Moten refer to as the "undercommon", that is, the actual, underground dimension of what is tenaciously and relentlessly produced as common: as something that stays behind, countering any partitive and ordering governmental logics; a "wild beyond", a "breach" or a "supplement" built upon the premise of fugitives lives "being together in homelessness" [22].

References

1. Agamben, G.: Homo Sacer. Il potere sovrano e la nuda vita. Bollati Boringhieri, Torino (1993)
2. Agier, M.: Gérer les indesiderables. Flammarion, Paris (2008)
3. Agier, M., Lacadet, C. (eds.): Un monde de camps. La Découverte, Paris (2014)
4. Agier, M., Bouagga, Y., Galisson, M., Hanappe, C., Pette, M., Wanesson, P.: The Jungle: Calais's Camps and Migrants. Polity Press, Cambridge (2018)
5. Anderlini, J.: Fitful circulations: unauthorized movements in the sicilian transit zone. Int. Migr. 1–13 (2022)
6. Aparna, K., Kramsh, O., Mahamed, Z., Deenen, I.: Lost Europe(s). Etnografia e ricerca qualitativa, 3/2017, pp. 435–452. Il Mulino, Bologna (2017)
7. Arendt, H.: The origins of Totalitarianism. Harcourt Brace, New York (1951)
8. Balibar, E.: Europe as Borderland. Environ. Plan. D Soc. Space **27**(2), 190–205 (2009)
9. Bigo, D.: Circulation et archipels de l'exception. Cultures et conflits, 68. L'Harmattan, Paris (2007)
10. Blight, D.W.: Why the underground railroad? In: Blight, D.W. (ed.) Passages to Freedom: The Underground Railroad in History and Memory, pp. 233–247. Harper & Co., New York (2001)
11. Bloch, E.: A Philosophy of the Future. Herder and Herder, New York (1970)
12. De Genova, N.: Migrant 'illegality' and deportability in everyday life. Annu. Rev. Anthropol. **31**, 419–447 (2002)
13. De Genova, N., Peutz, N. (eds.): The Deportation Regime: Sovereignty, Space, and the Freedom of Movement. Duke University Press, Durham (2010)
14. Deleuze, G., Guattari, F.: A Thousand Plateaus Capitalism and Schizophrenia. University of Minnesota Press, Minneapolis (1987)
15. Foner, E.: Gateway to Freedom: The Hidden History of the Underground Railroad. Norton & Co., New York (2015)
16. Foucault, M.: Sécurité, Territoire, Population, cours au Collège de France 1977–1978. Seuil/Gallimard, Paris (2004)
17. Garelli G., Tazzioli, M.: Containment beyond detention. The hotspot system and disrupted migration movements across Europe. Environ. Plann. D Soc. Space 1/19 (2018)
18. Gilmore, R.W.: Golden Gulag: Prisons, Surplus, Crisis, and Opposition in Globalizing California. University of California Press, Berkeley (2007)
19. Gordon, A.: The Hawthorn Archive. Letters form Utopian Margins. Fordham University Press, New York (2017)
20. Guareschi, M., Rahola, F.: Chi decide. Critica della ragione eccezionalista. Ombre Corte, Verona (2011)
21. Haraway, D., Goodeve, T.N.: How Like a Leaf: A Conversation with Donna J. Routledge, Haraway, New York (2000)
22. Harney, S., Moten, F.: The Undercommons: Fugitive Planning and Black Study. Minor Compositions, New York (2013)
23. Hartman, S.: Scenes of Subjection: Terror, Slavery, and Self-Making in Nineteenth-Century America. Oxford University Press, Cambridge MA (1997)
24. Heller, C., Pezzani, L.: Liquid traces: investigating the deaths of migrants at the EU's Maritime Frontier. In: De Genova, N. (ed.) The Borders of "Europe": Autonomy of Migration, Tactics of Bordering, pp. 95–119. Duke University Press, Durham, NC (2017)
25. Lafazani, O.: The significance of the insignificant: borders, urban space, everyday life. Antipode **53**(4) (2021)
26. Mbembe, A.: Necropolitics. Public Cult. **15**(1), 11–40 (2003)
27. Mezzadra, S., Neilson, B.: Border as Method, or, The Multiplication of Labor. Duke University Press, Durham, NC (2013)
28. Mezzadra, S., Neilson, B.: The Politics of Operations Excavating Contemporary Capitalism. Duke University Press, Durham, NC (2019)
29. Minca, C., Umek, D.: The new front of the Refugee crisis in the Balkans. Society & Space https://www.societyandspace.org/articles/the-new-front-of-the-refugee-crisis-in-the-balkans (2019)
30. Queirolo Palmas, L.: "Tra le macerie della Jungla di Calais. Reperti di una battaglia". Etnografia e Ricerca Qualitativa 3/2017, Il Mulino, Bologna (2017)
31. Queirolo Palmas, L., Rahola, F.: Underground Europe. Along Migrant Routes. Palgrave Mc Millan, New York (2022)
32. Rahola, F.: Zone definitivamente temporanee. I luoghi dell'umanità in eccesso. Ombre Corte, Verona (2003)
33. Rahola, F.: The space of camps: Towards a genealogy of places of internment in the present. In: Dal Lago, A., Palidda, S. (eds.) Conflict, Security and the Reshaping of Society. Routledge, London (2010)
34. Rahola, F.: "As we go along. Spazi, tempi e soggetti delle contro-condotte". Materiali Foucaultiani, 7/8 (2015)

35. Stierl, M.: Of migrant slaves and underground railroads: movement, containment, freedom. Am. Behav. Sci. 1–24 (2019)
36. Tazzioli, M.: Governing migrant mobility through mobility: containment and dispersal at the internal frontiers of Europe. Environ. Plann. C: Politics Space 3/19 (2020)
37. Tosatti, G.M.: New men's land. Storia e destino della Jungle di Calais. DeriveApprodi, Roma (2017)
38. Walters, W.: Border/Control. Eur. J. Soc. Theory **9** (2), 187–203 (2006)

Federico Rahola teaches Sociology of cultural processes and Theory of the image at the University of Genoa. He has taught and carried out research activities in France (Paris VII, MSH Nice) Finland (University of Jyvaskyla), Brazil (UFRJ and Puc Rio de Janeiro, Centro Josuè de Castro, Recife) and Canada (Concordia University, Montreal). He is the author of *Zone definitivamente temporanee. I luoghi dell'umanità in eccesso* (Verona 2003), *Chi decide? Critica della ragione eccezionalista* (Verona 2011, con M. Guareschi) e *Underground Europe. Lungo le rotte migranti* (Milano 2020, with L. Queirolo Palmas).

Inhabiting the Ecological Conversion: Experiments in Diavolution

Andrea Mubi Brighenti

Abstract

The paper reflects on the ecological transition from a political perspective. An attempt is made here to reassess Alex Langer's approach to ecology through the lens of 'conversion', examining the implications of his legacy for a new vision of citizenship. A suggestion is made to consider what could be the alternative to a 'symbolic' take on the ecological transition, retrieving a few indications from Peirce's pragmaticist philosophy.

Keywords

Ecological transition · Political agency · Pragmaticism · Social experimentation · Diavolution

1 Political Agency and the Act of Conversion

Since the mid-1980s, the environmental activist and co-founder of The Green Group in the European Parliament (then European Green Party), Langer [5–7], employed the expression 'ecological conversion' to indicate something more radical than a simple technological-engineering approach to the challenges raised by political ecology.[1] The word 'conversion' is, indeed, of religious origin, and evokes that existential *metánoia* whereby the subject's relation to the world is shaken to its foundations so as to engender a sudden, total change of the self *and* the world simultaneously.[2] Langer stressed that a technocratic approach to ecological problems was largely insufficient, insofar as it obliterated the issue of subjective *desire*: as he put it straightforwardly, 'we will only attain ecological conversion when it becomes desirable.'

Such desirability of conversion can emerge, in Langer's view, from a deep re-evaluation of the meaning of *boundaries* and *limits*: on the one hand, Langer's personal political commitment had led him to challenge a number of social boundaries that commonly separate closed groups, starting from the one between Italians and Germans in his native region, South Tyrol. Langer considered himself to be an 'ethnic objector', an 'ethnic traitor' and an 'ethnic deserter' who had rejected the official ethnic census policy that was—and still is

A. Mubi Brighenti (✉)
Department of Sociology, University of Trento, Trento, Italy
e-mail: andrea.brighenti@unitn.it

[1] Acknowledgments: I wish to thank Andrea Borsari for the invitation to join this volume, as well as two anonymous reviewers for the comments on a previous version.

[2] Significantly, in 2015 the expression 'ecological conversion' has been officially introduced into the Magisterium of the Church with the second encyclical by Pope Francis, *Laudato si'*. ([5]: §6, III).

A. Borsari et al. (eds.), *TEMPORARY: Citizenship, Architecture and City*, The City Project 4,
https://doi.org/10.1007/978-3-031-36667-3_5

—in place in the northernmost province of Italy.[3] Similarly, since the 1960s, Langer had in his political practice overcome the boundary between Catholicism and leftwing revolutionary activism; subsequently, in the course of the 1980s and 1990s, as a Member of the European Parliament, he actively pursued an ever closer cooperation between Western and Eastern Europe, fervently supporting the European enlargement, while never ceasing to critically challenge the mainstream understanding of Europe.[4]

On the other hand, since the 1980s, Langer was amongst those who most forcefully attracted the public attention towards the existence of clearly *impassable boundaries*, namely, the ecological *limits* of planet Earth. With respect to the ecological boundaries of the biosphere, Langer stressed, all humans are placed on the same side, that is, all are placed *on the inside* of such boundaries. Accordingly, the terrestrial boundaries of the biosphere come to define a condition of necessary 'interiority' of the humankind, where new measures must be set.[5] In this context, the *ecological conversion* advocated by Langer designates something that is at the same time both personal and social in extension, both attitudinal and practical in application, and remarkably trans-scalar in ambition, insofar as capable of working at different levels ranging from the very local to the very global.

2 The In-Between Time of Ecological Transition

While the act of *conversion* is a powerful one for the individual, and is associated to a radical existential *Umwertung*, that is, a re-evaluation of values that often spurs subsequent remarkable achievements, it is also plausible that modern society, in its technical and administrative complexity, might not be able to switch instantly into the ecological mode evoked by Langer and others. It may be the case, in other words, that a period of 'inhabitation' of the conversion itself must be envisaged. The phrase 'inhabit the conversion' might read as a paradox—in fact, I suggest, it is the most urgent societal skill to learn and train today. The terminology of 'transition' has been proposed to the same effect, and various European countries have indeed institutionalised such approach by creating 'ecological transition' ministries and programmes.[6] However, the current mainstream conception of transition is not only a watered-down version of conversion—it is also a *sterilized* conception where a remarkably technocratic-capitalist machinery (made of carbon bonds, eco-incentives etc.) has come to replace the humanistic understanding so passionately endorsed by Langer.

More amply, I think, the difference between transition and conversion can be excavated in terms of the question of inhabiting a shared city, which equates with the question of identifying a political *subjectivity* by bridging it to political *agency*. The gap between the radicalness of Langer's *conversion* and the sloppy, often ambivalent approach that we can see at play in current approaches to *transition*[7] is not by chance reminiscent of the classic twentieth-century gap between revolutionary and reformist political action. In this short reflection, I submit that,

[3] He personally bore the brunt of such a stance twice: first, in the 1980s, his application to move his post of high-school professor from Rome back to Bozen/Bolzano was frozen; subsequently, in the 1990s, his political bid to run for mayor of Bozen was struck down by a court decision—all of this because he lacked the 'declaration of linguistic belonging,' upon which the ethnic census is premised.

[4] As concerns the latter, see in particular Langer [8].

[5] Elsewhere, I have elaborated on the notion of interiority with reference to Elias Canetti's work, which raises very similar questions (Brighenti [2]).

[6] As of 2022, France, Spain and Italy have environment management ministries where the word 'transition' appears explicitly.

[7] A typical example is the widespread use of extremely toxic glyphosate-based herbicides in industrial agriculture. The dangers to human health posed by glyphosates are widely documented by medical studies. In 2017, Italy voted against glyphosates in the European Commission, but when the ban did not gather enough support, failed to implement a national law to prohibit them; in 2019, France passed a law to forbid these products, but waived the ban since the beginning of the pandemic in 2020 as a means to 'support' farmers; more generally, the large majority of European countries does not even have a phase-out calendar for glyphosates.

today, any real transformative process must address the blind spot of both revolution and reform. To do so, we must, in the first place, not loose sight of the fact that both revolution and reform fundamentally converge on the organisational plane. Whether through the continuous reformist path, or through the discontinuous revolutionary outbreak, a similar organizational dream can be seen at work throughout the spectrum of modernist political action [1]. That is why the issue of the *temporality* of the models for political ecology becomes pivotal, and the question I wish to address here concerns the possibility of letting emerge an ecologically-minded social temporality capable of playing with both continuity and discontinuity without falling prey to either techno-solutionist or politico-chimerical stances. Such is, arguably, the more profound meaning of the expression 'inhabiting the conversion.'

Alex Langer's work can be placed in the lineage of Ivan Illich's earlier analyses of the industrial society. Illich's problem resided in finding ways to conceive of an industrial society that does not end up being incapacitating or crippling towards its members. In this respect, both Langer and Illich can be considered as thinkers who embodied the pragmatist attitude. If, by pragmatism, we mean a philosophy that renounces principled decisions, and instead invites us to judge actions from their actual consequences, we must notice that such attitude cannot but be placed under the auspices of a philosophy of immanence—and this despite the fact that both Illich and Langer were religious believers. Yet only pragmatism, I think, can explain the emphasis these authors placed on 'tools': whereas Langer [7] invited us to 'equip ourselves for cohabitation', Illich [4] had called called for the elaboration of a wide set of 'tools for conviviality'. Illich, in particular, was keen on emphasising the paradox of the self-propelling nature of the institutional offer of goods in contemporary affluent societies: 'Our present institutions abridge basic human freedom for the sake of providing people with more institutional outputs' (*ibid.*, 25). That is why, he contended, 'people need new tools to work with rather than tools that 'work' for them' (*ibid.*, 23). The perverse effect of tools integrated into systems is that they increase the overall reliance and dependency, instead of fostering autonomy. Rather than surrounding ourselves with machines that do the work for us, Illich reasoned, we should rather focus on becoming the active utilizers of an enlarged set of instruments.

3 Learning to Train Desires

The criticism of the passivity of the consumer before the industrial-bureaucratic machinery was a defining feature of critical theories in the 1960s and 1970s. Just as in Illich, it can be found extensively among the Situationists, too. In my view, it is clear that, however important such a critique was at the time, today it no longer suffices in the task of envisaging a politics of emancipation. Indeed, there is now ample evidence that, with the 'activation' of consumers in digital capitalism, things have not necessarily improved: the coming of a 'postindustrial' society in the sense articulated by Illich, has not entailed the advent of ecological conversion—to the contrary, the hyperactivity of the new media and the performative turn on the workplace have led to enhanced ecological degradation. 'Activation', rather than 'passivity', is the key to understand the spoiling of natural and social goods as it unfolds today, with the appalling gravity it has touched in terms of climate change, pollution, loss of biodiversity. Economic theory must simply be turned upside down to make sense and match the current reality: it is not that the more people need fuel, the more it costs, but quite on the contrary, the more the cost of fuel rises, the more people drive. In other words, it is generalized activation that defines the systemic requirement of our age.

The pragmatist heritage remains of the essence, but a new politics of emancipation no longer resides in *activating* and *liberating* desires, rather, I suggest, it resides in *training* and *improving* them, so as to meet the requirements for ecological conversion—requirements that are in themselves not simply technical and pre-determined, but fundamentally ideational and non-deterministic. In this task, the kernel of the pragmatist stance can be retrieved in its sternly *anti-symbolic* stance. With reference to the threefold classification of signs elaborated by Charles Sanders Peirce, we should say that, in order to envisage ways to inhabit the ecological conversion, symbols *must give precedence* to *icons* and *indexes*. The reason is that iconicity and indexicality do not form codified systems ('habitual connections', in Peirce's parlance), but work operatively and tentatively—from the ground up, so to speak. Certainly, speaking of 'precedence' of the iconic and the indexical over the symbolic, does not mean that the symbolic should be simply discarded—for, as Peirce himself pointed out, all three forms are 'indispensable in all reasoning'.

The specific temporality called for by the task of 'inhabitation' is one that stands in opposition to the 'habitual', essentially *inertial* temporality that contradistinguishes the regime of the symbolic: in the domain of the symbolic, things become reasonable and predictable, tools come to be integrated and subsumed into systems.[8] That is why the temporal register of the symbolic is not part of the solution, but part of the problem. Icons and indexes offer a more promising terrain: icons correspond to the domain of likenesses and semblances—that ill-defined, informal field where similarities and affinities may appear without prearranged schemas; whereas indexes correspond to the active capacity to affect, act upon, and direct someone's attention. In short, with Peirce:

> There are three kinds of signs which are all indispensable in all reasoning; the first is the diagrammatic sign or icon, which exhibits a similarity or analogy to the subject of discourse; the second is the index, which like a pronoun demonstrative or relative, forces the attention to the particular object intended without describing it; the third [or symbol] is the general name or description which signifies its object by means of an association of ideas or habitual connection between the name and the character signified. (Peirce, *CP* §1.369).

We notice that, whereas the symbol largely operates in the domain of established significations and 'settled issues', both the icon and the index are experimental by nature. This is what makes them valuable to the present—for this interim period of inhabitation of the ecological conversion is an unsettled period almost by definition. Here, my idea is that we could put this time into the perspective of an open-ended enquiry to advance new modes of ecological existence. Inhabiting the ecological conversion would thus mean that we accept the 'experimental' stage we find ourselves at, and indeed turn into relentless experimenters ourselves. After all, nature is itself experimental, and science now tells us that we can no longer assume the general equilibrium postulate which underpinned classical ecological models of structuralist bent, but must face the reality of ecosystems evolving along trajectories—or, at the very least, in the process of transitioning across different equilibrium points.[9] Importantly, in my view, that does not mean that we should give up all criteria for *choosing* among possible different trajectories or courses of action and development—although it certainly does mean that, as we seek to repair the vessel, we are also on it, out on the open sea. In other words, although we cannot appeal to the 'preservation of ecological equilibrium' as sound reason to opt for one course of action over another one, we must not, just for that reason, give up the capacity to practically keep climate change and climate disaster apart from one another.

4 Experimenting with Transduction

Inhabiting the ecological conversion could thus practically mean that the direction and pace which we can hope to impart to societal dynamics must be figured out with the help of 'experiments.' These are not only scientific experiments in the classic sense, but coordinated collective arrangements where a series of forms of inquiry, discovery and practice can be

[8] Pierre Bourdieu had, in his sociology of *habitus*, pointed out something similar, although from a different perspective: according to Bourdieu, what habitus makes possible, and 'naturalised', is social domination.

[9] The latter thesis is, specifically, what the theory of tipping points suggests [9].

undertaken. Illich understood societal 'tools' as 'designed devices'; in line with this, the 'experiments' we need to conduct are, in the broadest sense, experiments *in design*—and, while we must remain alert to the fact that scale is a crucial political question of the present, given that scaling is never a neutral practice, we can also start small, were it not for the sense of empowerment that the very fact of concretely beginning an experiment conveys. Experiments are compressed, unique and outstanding moments: their temporality appears quite special, but also limited and bounded to a setting and a group of participants. That is why we must develop the ability to then translate such exceptional register into everyday existence—if we just were able to conduct one such experiment each day! Experiments tailored to inhabit the ecological conversion can be run in a relatively easy way by just staking out some dedicated time and space for them, bringing together people of different age, different classes, gender, orientations, preferences and so on, people with different skills, talents, passions and aspirations, so as to conduct intensive workshops and creative sessions where questions of inhabitation can be unpacked with care and in detail. They can involve anything as specific as imagining new recycling schemes, designing reusable portable containers, or as general as delineating new strategies for political mobilisation at the global level. Even experiments with *bad things* can be important: for instance, we can fruitfully frequent failed urban projects, learning form the ruins and the spoiled places that have been left behind by the machines of capitalist growth and of authoritarian regulation.

The possibilities offered by *conversion* as a mode of societal-ecological transformation can be clarified with reference to the notion of 'transduction' elaborated by the French philosopher Simondon [12]. It is necessary to read carefully the following extended passage by Simondon:

> By transduction we mean a physical, biological, mental, or social operation through which an activity propagates incrementally within a domain by basing this propagation on a structuration of the domain operated from one region to another: each structural region serves as a principle and model, as an initiator for constituting the following region, such that a modification thereby extends progressively throughout this structuring operation. The simplest image of the transductive operation is provided by the crystal, which, starting from a tiny germ, increases and extends following all the directions in its supersaturated mother liquor: each previously constituted molecular layer serves as the structuring basis for the layer in the process of forming; the result is an amplifying reticular structure. The transductive operation is an individuation in progress; within the physical domain, it can be effectuated in the simplest way via progressive iteration; but within more complex domains, like the domains of vital metastability or of the psychical problematic, it can advance with a constantly variable pace and extend into a domain of heterogeneity; there is transduction when there is an activity that starts from a being's structural and functional center and extends in various directions based on its center, as if multiple dimensions of the being appeared around this center; transduction is the correlative appearance of dimensions and structures within a being in a state of pre-individual tension, i.e. in a being which is more than unity and more than identity and which has not yet phase-shifted with respect to itself in multiple dimensions. The extreme terms attained by the transductive operation do not exist before this operation; its dynamism stems from the initial tension of the system of the heterogeneous being that phase-shifts and develops dimensions according to which it will be structured; it does not come from a tension between terms that will be attained and deposited at the extreme limits of transduction. Transduction can be a vital operation; in particular, it expresses the orientation of organic individuation; it can be a psychical operation and an effective logical procedure, although it is not at all limited to logical thought. In the domain of knowledge, it defines the veritable measure of invention, which is neither inductive nor deductive, but transductive, i.e. corresponds to a discovery of the dimensions according to which a problematic can be defined; it is an analogical operation, at least based on what is valid about this kind of operation. ([12]: 13–14)

Experiments in ecological transduction may be helpful for learning the art of breaking deadlocks and exiting from vicious circles. Most current societal and ecological problems are, effectively, problems of vicious circles, circles of dependencies (from loans, from chemicals etc.) that hamper the ecological conversion of society. In many cases, we are stuck into ecologically vicious cycles; and yet interestingly, as seen above, our social-ecological systems are also, at the same time, already in a condition of *metastability*—in other words, they are beyond equilibrium, and ripe for change. We are, accordingly, well placed for furthering new 'individuations' to come ('the transductive operation is an individuation in progress'). The fact that transduction begins from a 'centre' must be understood correctly, again in the perspective of a philosophy of immanence: the centre is not predetermined, to the contrary, the centre is *everywhere the transductive process takes hold* and begins a new structuration.

The ecological conversion cannot simply gesture towards sheer 'reconciliation' with nature, for such aim is still entirely premised upon the postulate of ecological equilibrium. Hence, the importance of what Simondon calls 'analogical operation': true analogy does not reduce difference, but amplifies it. It is not by chance, I think, that *analogy* is the same quality Peirce attributed to icons: neither individual nor general, iconicity proceeds through *singularities*, establishing flows of singularities across various objects in unforeseen ways. Again, the crucial idea is that likeness does not decrease difference, but to the contrary, it increases it. This suggests that transductive operations can be truly *inventive* to the extent that they are always effectuated through the *passage into a new dimension*. It is through such enrichment in dimensions that experiments with inhabiting the ecological conversion proceed most valuably. Transductive experiments must focus precisely on the emergence of those 'further dimensions' that enable us to reinterpret the problems previously posed in more reductionist way.

5 On Measure and Joy

In addition to analogical iconicity and its creative insight, the force of indexicality is to be factored into the equation, too. As considered above, indexes express the energetic aspect of semiotic relations: they are what makes a sign capable of directing someone's attention—if one wants, of making an actual change in the world. It is the indexical aspect that makes a sign truly living, acting, and compelling. For my part, some time ago I introduced the neologism 'diavolution' to indicate something akin to transduction in the domain of urban political relations. It may be expedient to report the passage where the ratio for the neologism is presented:

> What is diavolution? The cross-breed term replaces the Latin prefix *re-* with the Greek prefix *dià-*, which means 'through'. The Latin root *volvo, -ĕre*, which means 'to turn', remains. Accordingly, I propose to introduce the neologism 'diavolution' to address, for essentially descriptive purposes, the incessant activity of *going through* the problems that characterize the relationship between the nomic and the anomic. Diavolution is thus a movement that intersects the trajectories of these problems in multiple directions, or slantwise. From this perspective, diavolution can be described as a non-anomic way to avoid the nomic. Diavolution does not stand in opposition to revolution: it addresses a moment of desire which is present in many revolutions. Diavolution is not reformism, it is not withdrawal. It does not express an option for sub-optimal results or compromise, it does not aim at any paradigmatic settlement. Diavolution addresses those conceptual movements and practices whose outcomes are anything but certain because they are neither directed from a centre – as *emanations* – nor bound to an *éschaton-katéchon* dynamic. Diavolution is the immanent and a-centric presence of *volution*. It can only come about when a shift from the third to the second person takes place: *de te fabula narratur*. (Brighenti 2008: 797)

The shift to the 'second person' evoked in this passage resonates with the indexical force outlined by Peirce. Diavolution is, in other words, a mode of practical engagement, and can only make sense in terms of what Polanyi [11] famously called 'personal knowledge'. There is no a-priori way to tell which is the 'right', 'correct' and 'true' knowledge to develop, and this again calls for an experimental attitude grounded in pragmatism. It is not necessary that the pragmatist be an optimist (although, certainly, Langer *was* one); suffice that s/he remains a priori nonjudgemental, or, put differently, *sceptic*. The question of how to make ourselves 'at home' in the ecological conversion thus entails our remaining alert to the fact all 'homely' feeling is necessarily punctuated by the *unhomely* dimension of our condition. The disjointed state of the present lies in our being out-of-scale vis-à-vis the ecological requirements of existence. That is why our inhabitation of the ecological conversion is also inevitably going to be *uncomfortable*: the 'stain' cannot be removed. At the core of all ecological transductive experiments rests the issue of *measure*—again, with Illich, not only which type of tools we want to develop, but how many of them we should have:

> A convivial society should be designed to allow all its members the most autonomous action by means of tools least controlled by others. People feel joy, as opposed to mere pleasure, to the extent that their activities are creative; while the growth of tools beyond a certain point increases regimentation, dependence, exploitation, and impotence. ([4]: 33–34)

Tools have probably already grown beyond *that point*. In conclusion, it is not at all sure that we live in the era of 'Anthropocene': we will only be authorised to use that expression *only once* we have proven capable to avert the disaster we have been preparing for the living species, amongst which ourselves.

References

1. Brighenti, A.M.: Revolution and diavolution. What is the difference? *Crit. Sociol.* **34**(6), 787–802 (2008)
2. Brighenti, A.M.: Elias Canetti and Social Theory. The Bond of Creation. Bloomsbury, London (2023)
3. Francis (Pope): *Laudato si'*. Encyclical Letter (2015). Online at: https://www.vatican.va/content/francesco/en/encyclicals/documents/papa-francesco_20150524_enciclica-laudato-si.html, last accessed 16 Dec 2022
4. Illich, I.: Tools for Conviviality. Calder & Boyars, London (1973)
5. Langer, A.: Speech at Convegno Giovani di Assisi (1986). Online at: https://www.youtube.com/watch?v=b4ia9GYXJoQ, last accessed 16 Dec 2022
6. Langer, A.: We Will Only Attain Ecological Conversion When It Becomes Socially Desirable. Speech in Toblach/Dobbiaco, 10 Sept (1994a). Online at: https://www.alexanderlanger.org/en/279/1355, last accessed 16 Dec 2022
7. Langer, A.: Speech at Convegno Giovani di Assisi (1994b). Online at: https://www.alexanderlanger.org/en/279/1356, last accessed 16 Dec 2022
8. Langer, A.: L'Europa muore o rinasce a Sarajevo. June 25 (1995). Online at: https://www.alexanderlanger.org/de/34/163, last accessed 16 Dec 2022
9. Milkoreit, M., et al.: Defining tipping points for social-ecological systems scholarship. Environ. Res. Lett. **13**(033005), 1–12 (2018)
10. Peirce, C.S.: Collected Papers. 8 vols. Harvard University Press, Harvard (1932–1958)
11. Polanyi, M.: Personal Knowledge: Towards a Post-Critical Philosophy. University of Chicago Press, Chicago (1974[1958])
12. Simondon, G.: L'individuation à la lumière des notions de formes et d'information. Jérôme Million, Grenoble (2013 [1964–1989])

Andrea Mubi Brighenti is a social theorist based at the Department of Sociology, University of Trento, Italy. Research covers broadly space-power-and-society issues. His most recent book is *Elias Canetti and Social Theory. The Bond of Creation* (Bloomsbury, 2023).

Images of the People. Populus, Plethos, Plebs and Ethnos

Jacopo Galimberti

Abstract

Numerous scholars have explored the genealogy of concepts such as "citizen", "the people" and "sovereignty" from the perspective of the history of ideas and political sciences. In the last fifteen years, however, the academic interest in these analyses has been supplemented by lively debates that have occasionally reached the mainstream public arena in the Global North. In this context, social media platforms have played a major role, circulating innovative "images of the people" that tap in the history of political iconography. The essay focuses on the people as a performative entity whose contemporary visual instantiations deserve theoretical analysis. In particular, it combines the conceptual tools of political iconography with the framework delineated by a recent study about four meanings of the term "the people" (the people as "plebs", as "populus", as "ethnos" and as "plethos") in order to analyse brief case studies, and to suggest the need to conduct more in-depth analysis of visual negotiations of "the people" on social media.

Keyword

Citizenship · People · Sovereignty · Populism · Political iconography

In the context of the modern state, the notion of citizenship has often been associated with that of popular sovereignty. The discontinuous history of this conceptual bond commenced during the English Civil War, when "the citizen" contentiously supplanted "the subject" in parliamentarian factions, and the "will of the people", namely among the Levellers, began competing with the idea that sovereignty had a divine origin and resided in the monarch [1]. However, it was only in the late eighteenth and early nineteenth centuries—coinciding with the French and American revolutions, the first wave of decolonisation in Latin America and the rise of nationalism in Europe—that the rhetoric of the "appeals to the people", seen as the legitimate seat of sovereignty, acquired the mobilising force that it still has today. While in the latter part of the twentieth century the fervent invocations of "the people" as a usurped political subject was widely perceived as typifying the "populist" movements of the Global South, over the past fifteen years the claim of defending a deceived or betrayed people has regained prominence in the Global North, characterising discourses and visual practices that are embraced by parties across the entire political spectrum. Visual practices, in particular, have often played a

J. Galimberti (✉)
Department of Cultura del Progetto, IUAV, University of Venice, Venice, Italy
e-mail: Jacopo.galimberti@iuav.it

A. Borsari et al. (eds.), *TEMPORARY: Citizenship, Architecture and City*, The City Project 4,
https://doi.org/10.1007/978-3-031-36667-3_6

prominent role in the shaping of a construct as complex as "popular sovereignty". Indeed, while the portrait of the person of the King and his attributes would have been known to the population (for example, through coins), and the King's supposed "divine right" could rely on a longstanding religious iconography depicting divine interventions in the terrestrial world, imagining "the people" as a collective entity capable of expressing a will and incarnating something superior to a government required a suspension of disbelief and the forging of an unprecedented visual repertoire. Since the late eighteenth century, painters, sculptors, caricaturists, and architects, etc. have devised a great quantity of both novel and familiar motifs that have aimed to render the fiction of popular sovereignty not only recognisable but also credible. Oftentimes, these reused forms have been borrowed from personifications informing absolute sovereignty [2]. Yet for all the power inherent to this imagery, "the people" (meaning the subject exerting sovereignty) cannot be anything but performative. This point has been explored by several contemporary political theorists. Pierre Rosanvallon has contended that "'the people' does not pre-exist the fact that it is invoked", an assertion echoed by Judith Butler in one of her most recent books, maintaining that in order to achieve a public existence, collective subjects must constantly be performed lest they become unthinkable [3]. It can be said that every visual iteration of "the people" contributes to the shaping of the legal and political contours of this subject, as well as of the individual units that compose it. In this case, the citizens.

Throughout the history of art and architecture there are masterpieces that engage directly with the depiction of sovereignty and its bearers—from Jacques-Louis David's *The Tennis Court Oath*, whose subtle visualisation of power has been analysed by cultural theorist Stefan Jonsson, to "spheric architecture", the political agenda of which has been delineated by art historian Susanne von Falkenhausen [4]. Rather than supplementing these studies with further analyses of major artworks depicting popular sovereignty, this chapter aims to establish a dialogue with, and to offer a visual counterpart to, the conceptual grid that has recently been crafted by Gérard Bras in an attempt to examine the different meanings of the term "the people" [5]. After a short summary clarifying Bras' theoretical findings, the chapter will explore the four overlapping yet distinct connotations that this term has acquired from a visual perspective. The investigation will revolve around contemporary images sourced predominantly from social media platforms. This focus on visual tropes has a twofold rationale. First, it intends to demonstrate the contemporary relevance of this type of analysis, too often confined to the history of art. Second, it aims to complicate Bras' framework, which is based on words, whereas contemporary appeals to the people are often mediated via visually rich social media.

1 Four Types of People

In his 2018 book, *Les voies du people*, Gérard Bras delineated a conceptual history of the notion of people (*peuple*) and the multifaceted nuances this word has carried with it over the past two and half centuries. Largely complementing the exploration of "historical semantics" conducted by Reinhart Koselleck's research group, Bras' volume has identified and elucidated, among other things, four main acceptations: "the people" as *populus*, as *plethos*, as *plebs* and as *ethnos* [6]. As *populus*, the people are regarded as the legitimate holder of sovereignty, as an abstract juridical notion that can identify the citizens of a country, but also a numerically limited group; an example is the case of male-only suffrages or of suffrages, common in Europe before World War I, that were virtually restricted to the middle and upper classes (as in "household suffrages" and suffrages excluding illiterate citizens or those whose income taxes fall below a certain level). The people as *plethos* frames the people as the many; this is the vast majority of citizens who are often seen participating in some form of collective action. The core idea behind this type of "people" is that masses matter in politics, and that a government sustained by the elite may be legal but cannot be considered fully legitimate and cannot speak on behalf of all of the people. In the twentieth and twenty-first centuries, the main rationale behind attempts to stage the people as *plethos* has been to generate a set of images encoding a supportive or threatening message towards an authoritative body. The idea of the people as *plebs* may imply that the people must be *plethos*, but it cannot be reduced to it completely, as the notion of *plebs* characterises the people in social terms. It is not (or it is not only) their quantity that provides the citizens/denizens with authority, but rather it is what they lack—be it the means of production, land, incomes, civil rights, etc. The exclusion from the political arena of the people as *plebs* is often framed as a usurpation, and the debased status of this "people" is cast as a noble condition awaiting vindication. The people as *ethnos* indicates a population that supposedly shares some defining traits: a language, a "race", cultural mores, geographic origins, etc.

Bras' conceptual grid is largely predicated on texts, but it can also be operationalised by observing how some of its notions highlight the visual strategies of contemporary politics and its manifold "appeals to the people".

1. The people as *populus*. In 2015 Pablo Iglesias Turrión, the leader of Podemos (a Spanish leftist party), announced his candidacy for Prime Minister. The backdrop for Iglesias' press conference (Fig. 1) featured a work by Jose Delgado Periñán.

   https://nomoi.hypotheses.org/263

Fig. 1 Pablo Iglesias presenting the list of Podemos' candidates and announcing his candidacy for prime minister at the next general elections

This was a drawing titled *Todos los demas* (everybody else), published in *Multitudes*, an artist book authored by Jose Delgado Periñán himself as well as Miguel F. Campón. Through its carefully choreographed communication, Podemos solicited comparisons with the most studied image of political philosophy, the frontispiece of *Leviathan* (1651), for which Thomas Hobbes had Abraham Bosse depict a stately character to conjure "the people" [7]. In Hobbes' books *Leviathan* and *De Cive* the notions of "multitude" and "people" bear specific meanings. In short, a "multitude" of disperse and ruthless individuals with no rights and laws need to reduce their will to one, submitting themselves to a sovereign who will enforce order, allowing them to be a "people", a *populus*, writes Hobbes in *De Cive*. This transformation takes place in exchange for the transfer of individuals' rights to the monarch in order to use that power to preserve their lives—an abdication necessitated by the condition of humankind. Indeed, before the sovereign receives the mandate to rule, there exists only a perpetual war, according to Hobbes, a brutal "state of nature". The frontispiece of *Leviathan* does not depict a colossal monarch, but rather the personification of the "body politic", whose arms and torso are constructed from a number of individual figures. By submitting to the sovereign, these are no longer the "multitude", but rather the *populus*.

Podemos' body politics is both similar to, and divergent from, that gracing the frontispiece of *Leviathan*. While its iconographic model is akin to that devised by Bosse –to use Dario Gamboni's terminology, a "composite image" (in other words, a human body containing smaller human bodies), Iglesias is located below the people, suggesting the latter's primacy [8]. His body does not, or at least does not yet, subsume the *populus* as in *Leviathan*, but it seems willing to receive the descending crowd in what is cast as an ongoing process. Iglesias will grant them unity, the photograph suggests, yet this unity is already immanent in the throng's shape, which is vaguely reminiscent of that of Spain itself, further grounding the *populus* in the "Nation". Another difference between Hobbes' dialectic of the monarch and citizens can be detected in how the individuals are represented. Whereas Bosse crafted a differentiated crowd where each person has distinctive features, Podemos embraced a faceless crowd of anonymous citizens. This iconography tends to intimate, among other things, the presence of an absence in the contemporary democratic public space [9]. In particular, the faceless crowd refers to the demands of the Spanish anti-austerity movement (also known as "the 15-M Movement"), from which the party emerged in 2011–2012.

If, following Bras' analyses, the term *populus* designates that the people are entitled to deliberate over important matters in the public arena, it is possible to argue that Podemos casts the *populus* as the assembly of the excluded, of the marginalised, as the "everyone else" of the nation who should rally behind Iglesias. The idea of the *populus* as the disenfranchised introduces another meaning of the word "people": the people as the *plethos*, as the vast majority of the population.

(2) The people as *plethos*. What took place in the halls of the United States Capitol Building on 6 January 2021 can be interpreted as a new praxis in the repertoire of contemporary protests: the symbolic (but not peaceful) intrusion into a government building by individuals who consider themselves to be "the true people" because they are *plethos* and, as such, solely legitimate. On a smaller scale but with similar symbolic goals, a group of yellow vests had used a forklift as a battering ram to break open the gate of a Ministry in Paris on 5 January 2019, two years before the assault on the Capital. By the same token, German marches against the government's policies over COVID-19 had repeatedly tried to penetrate the Reichstag (the German parliament) and in November 2020 a handful of protesters gained entry. More recently, hundreds of protesters demanding the resignation of President Gotabaya Rajapaksa rampaged through his official residence in the Sri Lankan capital Colombo, and took a swim in its luxurious pool.

Regardless of their political connotation and the level of violence involved, these types of actions reveal a paradox inherent to popular sovereignty. By forcibly imposing the citizens' physical presence into places where citizens are not normally permitted, but by definition are present in the form of their delegates, protesters highlight the problem of limiting or ousting a government that derives its authority directly from the people for whom it has claimed the right to speak. One is poised to wonder whether popular sovereignty implies that the *populus* is ultimately sovereign only on election day. In any case, what is novel about this protest repertoire is its symbolic dimension, which distinguishes it from insurrectionary days where the occupation of a government building was integral to an uprising aiming to seize power. However, a cursory look at the depictions of the people as *plethos* in the context of the storming of the Tuileries in 1792 and 1848 shows two recurrent motifs connecting recent protests to past revolts: the performance of actions associated with the naked body in the majestic halls of official buildings (such as urinating or swimming) and the presence of "low" and unclean body parts (typically a bare foot or the buttocks) ostentatiously placed on the symbols of power [9].

While resisting the fiction of "political representation", the alleged justification of the protesters' actions inevitably enacts another fiction: that a small group of citizens can temporarily be the *populus*, or at least its bravest emissaries can. In today's technological landscape, this change of scale is easily recognisable: sharing images on social media that "go viral", or protestors livestreaming their actions in a way that fosters the illusion that they are backed by more supporters than they truly are. It is no coincidence that the protester (Fig. 2) provocatively placing his shoe on Nancy Pelosi's desk holds his smartphone in his hand. In Washington, DC, in particular, the rapid and immense mediatisation of the attack allowed a few hundred citizens to see themselves as a *plethos*, as a throng whose political and moral legitimacy—supposedly superior to the mere legality of the newly elected government—was granted by its sheer magnitude.

Fig. 2 One of the protesters, Richard Barnett, during the attack on the United States Capitol

https://edition.cnn.com/2021/04/26/politics/nancy-pelosi-richard-barnett-capitol-hill-riot/index.html

In Paris as much as in Washington, in 1792 just as in 2021, the optical clash between humble, "low" body parts and authoritative symbols of the polity encodes a further irreconcilable tension in the legal architecture of popular sovereignty. This tension speaks to the perception of the people as *plethos* as having "two bodies", to paraphrase Ernst Kantorowicz's celebrated book *The King's Two Bodies* [10]. Indeed, the word "people" occasionally conflates the abstract *populus*, the juridical subject endowed with sovereignty, with the vast and very concrete *plebs*: the lower classes.

(3) The people as *plebs*. According to Bras, it is only with the philosophy of Jean-Jacques Rousseau that the *populus* can no longer be easily dissociated from the people as *plebs* [11]. While the identification of the working classes with the "best part" of the *populus* has long characterised the leftist parties' rhetoric, today, at a time when the category of "working-class aesthetics" has been largely evinced from the study of visual culture, it is the far-right parties that strive to perform and re-signify aspects of the working-class and "plebeian" ethos in their online communication [12]. Consider the image (Fig. 3) posted on Facebook by Lega (a far-right party that was part of the government) shortly after the 2019 European elections. The man in the picture, Matteo Salvini, was then the Minister of the Interior and the leader of Lega, which had obtained 34% of the vote at that election. The sheet he holds in his hands displays a text added after the photograph was taken that reads, "First party in Italy. Thank you". There are also some crucial messages located behind Salvini, however, which are more visual in nature. Indeed, his library/display cabinet in the ministry features more than 19 items that present a carefully calibrated collection of symbols designed to instantiate Salvini's putative political culture and personal life, from a book on Heinrich Himmler (one of the most prominent figures of the Nazi party) ("16") to Christian icons ("11"), a cap with Trump's slogan "Make America Great Again" ("9"), mundane objects such as a selfie stick ("15") and an autographed photo of an AC Milan football player ("1"). The links between the objects in this crammed space and the *plebs* requires a short deviation from the main discussion.

Fig. 3 Matteo Salvini thanking his electors after the European Elections of 2019

https://www.repubblica.it/politica/2019/05/27/news/elezioni_europee_foto_salvini_messina-227322060/

The hiatus between an art *for* and an art *of* the lower classes was a leitmotif of the modern movement in the 1920s and 1930s. At least two broad attitudes can be discerned. On the one hand, some modern architects and designers laboured to disseminate "true" concepts of proletarian dwelling types, as they believed that what proletarians understood by "personal taste" was nothing more than the visual expression of the ethos of other social classes. The defenders of this rather essentialist view of proletarian culture often championed sobriety and practicality as its most desirable features. On the other hand were intellectuals who objected to this notion: because of the bareness of their working environment, proletarians needed sentimental surroundings at home. Thus, designers and architects needed to refrain from restoring "order" in the cramped space of working-class households. Instead, they should encourage their proclivity for lurid fabrics, "romantic" nooks, mawkish trinkets and memorabilia [13].

The motley assemblage of Salvini's library/display cabinet, which was not located in a private space but rather in his office at the ministry, mimics the plethora of "gaudy" knick-knacks and Catholic imagery typical of lower-class Italian households. His self-fashioning plays on the combination of an allegedly average Italian citizen, a "common man" informed by apolitical pop culture and right-wing icons, thereby merging "extreme and mainstream" [14]. By putting "low" objects in the luxurious halls of the government office, Salvini performs a gesture that is no different to that of the protester placing his shoe on Pelosi's desk. Yet Salvini had been appointed as Minister of the Interior, therefore his presence in a government building is completely legitimate. Through this widely circulated and carefully staged imagery, Salvini attempted to fuse the *populus*, embodied by Salvini himself as an elected minister in his office, and the *plebs*, whose ethos is conjured by the objects on display and by Salvini as their proud owner. The ostentatious presence of Catholic icons carries not only social connotations but political ones as well, asserting that Salvini's "people" are not only *plebs* but also *ethnos*; that is, inextricably rooted in a specific tradition and history.

4. The people as *ethnos*. Throughout history, the connections between citizenship and *ethnos* have varied considerably. In ancient Greece, for example, in order become a citizen of a polis, to belong to the Greek *ethnos*, one had to be a free-born man and speak a dialect of Greek. In ancient Rome, what is here called *ethnos* only had a tenuous connection with citizenship (*civitas* in Latin), whose obtention often stemmed from a political decision dictated by strategic and economic circumstances independent from the people's languages and mores. With the rise of nationalisms and racist theories in the nineteenth century, it became common to "naturalise" nationalities and to use terms such as "Italian race" or "Spanish race". Since World War II, these references to race have slowly lost their popularity in Europe, and contemporary far-right parties tend to anchor the *ethnos* (and the claims of those belonging to it) to "traditions", "culture" and "civilisation".

One of the most recent examples of this transition is offered by Vox, a far-right Spanish party that is clearly nostalgic of the Francoist dictatorship. Founded in 2013, Vox has always been grounded in a media landscape marked by the heavy use of social media platforms and their specific forms of communication and affordances. Vox's rather monolithic idea of the Spanish people as *ethnos* can be exemplified in an image (Fig. 4) that the party posted on its main Instagram account on 8 December 2019, the day of the Immaculate Conception. The "patron saint of the Spanish Infantry", read the comment accompanying the picture. The image depicts *The Miracle of Empel* (2015), a work by Augusto Ferrer-Dalmau Nieto, a contemporary Spanish hyperrealist artist who specialises in historical military paintings. This was part of a series of Instagram posts in which Vox referenced the *Siglo de Oro* (The Spanish Golden Age)—another being a post featuring Bartolomé Esteban Murillo's painting *The Adoration of the Shepherds* (1668) associated with the Spanish flag—this work refers to a battle unexpectedly won by the Spanish troops in the Netherlands in 1585. The pious religious feelings conveyed by the scene and its sources in the *Siglo de Oro* reconnects the party to a supposedly "authentic" Spanish Catholic tradition and a glorious cultural legacy. By the same token, the Instagram post indirectly conjures the Christian fervour

Fig. 4 Augusto Ferrer-Dalmau, El milagro de Empel (2015), posted on Vox's Instagram account on 8 December 2019

typifying the military epos of the Reconquista (against Muslims) and, not least, the iconography of Francoism [15]. Indeed, Franco's regime frequently made recourse to Spanish Baroque art in order to foreground a royal lineage (Franco was *caudillo* "by grace of God"), to praise the rigid morality of the Counter-Reformation and, crucially, anti-Communism, with the Reconquista alluding to the defeat of the "Reds" during the Spanish Civil War [16].

https://www.lavozdigital.es/cadiz/lvdi-inmaculada-origen-y-curiosidades-esta-celebracion-201912071703_noticia.html

Through this image, Vox contributed to the performance of the Spanish people as *ethnos*, as a relatively homogenous group sharing an illustrious past, an immutable culture, a religion and, most importantly, the same enemies—be them Muslims, leftists or the LGBTQ + community. Noting the fact that many Spanish citizens are atheist or do not identify with the values of Francoism would be to miss the point raised by Vox's pointed recourse to a mixture of pop culture and Baroque imagery. Indeed, it would directly contrast with Vox's mythical idea of the people as *ethnos* with another, not necessarily compatible with the idea of people as *populus*. While citizens are safeguarded by the *Rechtsstaat* (or "rule of law") in the latter, the former tends to see them as members of a sacred, unchangeable spiritual community that must be defended from both foreign and domestic enemies, regardless of their nationalities.

2 New People, Citizens and Images

At a time when appeals to the people have once again become common currency in the Global North and the contentious issue of sovereignty has regained momentum, notably among Brexiteers and sovereigntists more generally, meditations on citizenship need to engage with how the political "contours" of the people are negotiated and enacted through digital images on social media platforms. In this essay, the categories delineated by Gérard Bras for the study of the notion of the people as a sovereign political subject have been used to explore social media images and to delineate tensions immanent to the iconography of the people. In particular, the analysis has focused on four acceptations of "the people", the people as *populus*, as *plethos*, as *plebs* and as *ethnos*. This type of iconological investigation aims to complement conceptual history and the history of art, bringing these disciplines closer together following the path initiated by the *Politische Ikonologie* [17]. Yet it also hopes to foster debates among designers, artists and social media managers willing to shape campaigns that counter "toxic" narratives surrounding sovereignty and peoplehood. Indeed, it is rather unlikely that the supposedly impulsive feelings elicited by "populists" can simply be tamed by providing "rational arguments" and "facts". Reason cannot entirely govern "affects", to adopt Baruch Spinoza's terms and his implicit "theory of ideology" [18]. Precisely because of its largely fictional nature, the will of the people cannot be separated from the ability to produce cogent and emotionally charged depictions of it. In order to produce new affects and shape innovative frameworks for the notion of citizenship, we need, now more than ever, novel and powerful images of the people.

References

1. Morgan, E.: Inventing the People: The Rise of Popular Sovereignty in England and America. W.W. Norton, New York (1989)
2. Manow, P.: In the King's Shadow. Wiley, New York (2014)
3. Rosanvallon, P.: Le peuple introuvable: histoire de la représentation démocratique en France. Gallimard, Paris (2010), p. 24; Butler, J.: Notes toward a Performative Theory of Assembly. Harvard University Press, Cambridge, Mass (2015)
4. von Falkenhausen, S.: KugelbauVisionen: Kulturgeschichte einer Bauform von der Französischen Revolution bis zum. Transcript Verlag, Bielefeld (2008); Jonsson, S.: Crowds and Democracy: The Idea and Image of the Masses from Revolution to Fascism. Columbia University Press, New York (2013)
5. Gérard, B.: Les voies du peuple: éléments d'une histoire conceptuelle. Éditions Amsterdam, Paris (2018)

6. Koselleck, R., Gschnitzer, F., Werner, K.F., Schönemann, B.: Volk, Nation, Nationalismus, Masse. In: Koselleck, R., Conze, W., Brunner, O. (eds.) Geschichtliche Grundbegriffe. Historisches Lexikon zur politisch-sozialen Sprache in Deutschland, vol. 7, Klett-Cotta, Stuttgart (1992), pp. 141–431. Bras focuses particularly, but not exclusively, on the French-speaking world
7. For this image, see Bredekamp, H.: Thomas Hobbes - Der Leviathan: Das Urbild des modernen Staates und seine Gegenbilder. 1651–2001, De Gruyter, Berlin (2001)
8. Gamboni, D.: Composing the Body Politic. Composite Images and Political Representation, 1651–2004. In: Latour, B., Weibel, P. (eds.) Making Things Public. Atmospheres of Democracy, pp. 108–119. MIT Press, Cambridge, Mass (2005)
9. See Galimberti, J.: Democracy and masks. Towards an iconology of the faceless crowd. Eur. J. Creative Pract. Cities Landscapes **4**(1), 211–235 (2021)
10. Kantorowicz, E.H.: The King's Two Bodies: A Study in Mediaeval Political Theology. Princeton University Press, Princeton (2016)
11. Gérard, B.: Les voies du peuple: éléments d'une histoire conceptuelle, p. 39. Éditions Amsterdam, Paris (2018). For Rousseau, see also Frank, J.A.: The Democratic Sublime: On Aesthetics and Popular Assembly. Oxford University Press, New York (2021)
12. For this topic, see the forthcoming issue of the Oxford Art Journal, devoted to the issue of "Art and class today" (1960–2020)
13. For examples of this dichotomy in "Red Vienna", see for instance, Blau, E.: The Architecture of Red Vienna, 1919–1934, pp. 192–196 and 378–380. The MIT Press, Cambridge, Mass. (1999)
14. Miller-Idriss, C.: The Extreme Gone Mainstream: Commercialization and Far Right Youth Culture in Germany. Princeton University Press, Princeton (2020)
15. Esteve-Del-Valle M., Costa López, J.: Reconquest 2.0: the Spanish far right and the mobilization of historical memory during the 2019 elections. Eur. Polit. Soc., published online on 7 Apr 2022 (2022)
16. For this topic, see Galimberti, J.: Individuals against Individualism. Art Collectives in Western Europe (1956–1969). Liverpool University Press, Liverpool (2017)
17. Fleckner, U., Warnke, M., Ziegler, H., (eds.): Handbuch der politischen Ikonographie. Beck, Munich (2011)
18. Del Lucchese, F.: Tumult and Indignation. Conflict, Power, and Multitude in Machiavelli and Spinoza. Continuum, London (2011)

Jacopo Galimberti is an art historian and assistant professor (with tenure track) at IUAV (Venice). His articles have appeared in journals such as *Art History*, *The Art Bulletin*, *The Oxford Art Journal* and *Grey Room*. He is the author of *Individuals against Individualism. Western European Art Collectives (1956–1969)* (Liverpool University Press, 2017), of *Détournement & Kitsch. Die Postkarten von Hans Peter Zimmer/Les cartes postales de Hans Peter Zimmer* (Les Presses Universitaires de Nanterre, 2021) and of *Images of Class: Operaismo, Autonomia and the Visual Arts (1962–1988)* (Verso Books, 2022).

Making the Temporary Become Chronic: The Transit City

Pierpaolo Ascari

Abstract

When the French authorities resolved to dismantle the Parisian bidoville, the new dispositif they resorted to was that of transit cities, a programmatically provisional solution that still today, however, awaits to be completely overcome. In Nanterre, for example, the definitive reconversion of the cité des Potagers into a district of rent-controlled housing will only be completed in 2025, but in the meantime the entire history of these complexes will have resulted in the chronicisation of an "unfinished urbanity", as Abdelmalek Sayad defined it, determined on the border between the dimension of the temporary and the more implicit one of the class, gender and race hierarchies that the reference to the temporary allows to eternalise.

Keywords

Temporary · Transit · Détournement

P. Ascari (✉)
Department of Architecture, University of Bologna, Bologna, Italy
e-mail: pierpaolo.ascari@unibo.it

1 Method Premise: Against Ventriloquism

The perspective in which I would like to try to make my contribution is that for which citizenship understood as a material condition of living in the city, can be considered an act of citizenship in the same way as conflicts, strikes, marches or any other kind of pressure that is more clearly mobilised at the confines of the formal statute of citizenship. By this I do not mean at all to deny the importance of struggles and their organisation, which remain essential and peculiar, but rather to support the thesis that their importance and this peculiarity derive from the fact that the various subjects involved in the forms of claiming citizenship are constituted first of all in the material (and sometimes illegal) struggle for existence. To exist and survive, for a historically not insignificant part of citizenship, has meant and still means exercising a right not yet recognised by the legal systems and therefore performing an act of performative citizenship.

In conclusion, we will mention how this extension from the struggle for rights to survival practices, in order to be theoretically plausible, involves a rereading of the text to which the reflection on performative citizenship is usually traced [1], which is that of the William James Lectures held by John Austin at Harvard University in 1955, which later became famous under the title *How to do things with words.*

A. Borsari et al. (eds.), *TEMPORARY: Citizenship, Architecture and City*, The City Project 4,
https://doi.org/10.1007/978-3-031-36667-3_7

A final premise to the analysis of the case that will allow me to set up this reasoning concerns a second misunderstanding in which I would be sorry to run into, that for which to qualify forms of life from the perspective of citizenship would be something politically implicit and unaware, that a certain sociology has identified with the *infrapolitical* of the gestures and actions carried out by those who would not be fully able to understand the reasons and the scope. Because what substantiates this significance, Donna Haraway would say, is always a *ventriloquy*, an attribution of meaning with which reality remains at the disposal of its interpreters, even the most benevolent [2].

In one of the evolutions of the case to which I will have to refer, that of the transformation of the Parisian *bidonville* into the transit city (*cité de transit*), the persistence of this epistemological pose appeared in all evidence in the autumn of 2005, when there were not a few or the worst to identify the protagonists of the riots in the banlieue with the "rebels without a cause" of the homonymous film by Nicholas Ray. As for the character played by James Dean, it was therefore necessary to postulate a cause that referred to the relationship with the parental culture, to a story of fathers and sons that once recognised could have made the real difference between the wrongs and the reasons. It still happens today that we come across the speculative horizon of these mythologies and their attempt to redeem violence and youthful vandalism, for example, by placing them in terms of an "informal register of dissent" [3]. What seems to me to emerge, however, is how to operate on the border between survival and formal citizenship, if only in reference to the case I dealt with, is an absolutely formal (in the sense of formally determined and transmissible) register of experimentation: not always profitable, sometimes dramatic, but still an alternative to the postulate of alleged deficiencies that would imply the need for ventriloquism.

2 Coal in the Bathtub

The case in point, as I mentioned, is the dismantling of the Parisian *bidonville* (but not only) and of the instrument through which it was decided to implement it: the *cité de transit*. A dispositif that knows its period of greatest diffusion at the beginning of the seventies, but which inherits some fundamental characteristics from the other instruments with which the French administration has already faced the emergency of Algerian, Moroccan, Portuguese, Spanish and internal immigration, especially of labourers, in the previous decade [4]. *Ethnic foyers, emergency city, promotional city*: these are the names that now correspond above all to the new urgency to govern a migratory flow from Algeria that the War of Independence (1954–1962) made at the same time more voluminous, composite and turbulent. Voluminous and composite, because now it is not only the workers who immigrate, but also their families, since the war would have forced them to flee anyway and therefore proves to be a valid reason for derogating from the tradition that required men not to involve women and children in their migration project. Turbulent, because now the members of the National Liberation Front can camouflage among the migrants, infest the shanty town and make proselytes. The emergency city and the promotional city therefore already respond to a programme of a panoptic order that only since 1967 has been specified in the corrective and officially preparatory vocation of the transit city.

At this point, the transit will no longer be understood as a temporary presence of the immigrant inside the hexagon, but as the passage of a defective population to a different way of living and inhabiting the city, finally conforming to customs more civilised than the former colonisers. From now on, the transit cities will have the task of taking charge of the entire lifestyle of the families who will be relocated there, since if the shanty town represented "leprosy", "cancer" or "gangrene" which risked infecting the urban organism, its inhabitants—in the absence of a transitory accompaniment to civilisation—can infect the social body and therefore require adequate care. The frequent medical and surgical metaphor, then, is flanked by the grey literature of decrees and ordinances with the usual resources of the *psy* discourse, according to which to distinguish these families would be, for example, "poverty in phantasmatic elaboration", a character that would prevent them from imagining a future other than the past misery in which they lived [5].

Thus, in the new agglomerations, all the families whose "access to permanent accommodation is not feasible without educational action" will have to end up for a limited period, intended to correct not only the atavisms attributable to their geographical origin, but also the bad habits that in the meantime they will certainly have acquired in the defective habitat of the slum. We must act, provide individuals who come from the slum with a "comfortable and reassuring" solution in order to "tame them, encourage them, familiarise them with the first needs of modern life and coexistence". Centralised lifts, hot water and waste bins will come later, when Algerians will have assimilated the first advantages of metropolitan comfort.

Especially since when the elevators are there they fill up with graffiti or urine, because "it is known that these people break everything" [6]. This is reported by a pioneer of urban anthropology, Colette Pétonnet, in an attempt to synthesise the point of view that institutions aspired to naturalise through the elaboration of their preparatory mission. A mission that at this point can be fully understood, perhaps, by relying on Colette Guillaumin's reflections on the notion of *mark*, with which it enters into considerable resonance.

According to Guillaumin, we are dealing with marks when a consequence of social relations tends to stand out as their origin. Within this inversion, the very idea that there may be individuals naturally inclined to break everything, for example, serves to make permanent the validity of certain class, race and gender hierarchies through the perpetuation of their effects. These hierarchies are therefore considered a product of the internal traits of those who suffer from their reiteration, the result of an "endogenous determinism" that the analysis of the mark system has the task of deconstructing [7]. And it would seem that Guillaumin identified the dispositif that governs the transit city, which according to the Vivien law of 10 July 1970 will be built to be disassembled within 10 years (i.e. with inexpensive materials and no services) and that every two years should have foreseen the rotation of the inhabitants.

Indeed, twenty years later, when both the residents and the buildings are still there [8], the preparatory habitat will finally be revealed for what it is: *an emergency dispositif that serves to rationalise exclusion through the chronicisation of the temporary*. And in the meantime it will be the alleged "endogenous determinism" of the inhabitants to have expressed itself in the forms of making-bidonville (*bidonvilisation*), that is to say in the presence of mud everywhere and in the soiled walls, in the frequent fires, in the breaking of drains. Because these people break everything, they said, because they are not used to comfort and collective housing, because they do not know how to use them: it will still be the old example of "coal in the bathtub", as for the former slaves of which Francis Scott Fitzgerald tells in *Diamond as big as the Ritz*, to prove how much it is preferable for families housed in the *cité* not to move elsewhere, to apartments and neighbourhoods that, precisely because they are better served and more in step with modernity, would conflict with their natural propensity to use the bathtub to store coal [5].

3 An Almost Surrealist Détournement

But the making-bidonville does not limit itself to naming the course of the alleged pathologies that the mark system attributes to the endogenous determinism of individuals unnecessarily subjected to the corrective enterprise of transit. Mentioning it in a potentially different perspective is Abdelmalek Sayad in his ethnographic work on the Nanterre slum, where entrepreneurs have found ways to stabilise immigrant labour, "sleep merchants" have had more sleep to sell, and the police have been able to control the homeless population more easily. If the slum is presented as "a continuous creation", says Sayad, it is above all because it constitutes the always dynamic balance point between a multiplicity of different interests. Nevertheless, he adds, it is only to the skin of those who live in it that mud sticks, the "mark of an infra-urbanity" that cannot entirely refer to either the city or the countryside, "because in the slum neither can assert itself as such", thus determining a condition of "unfinished urbanity" that outrages the metropolis, its aesthetics and its values [9]. And the mud becomes the main reason for recognition and ridicule when the inhabitants of the Nanterre shantytown go to the city fountain to stock up on water, but alongside a second character highlighted by Sayad which in the development of our reasoning could turn out to be decisive. I quote in full, because it seems to me the crucial passage of the investigation: "There is something moving in hearing today the former inhabitants of the slum speak with the greatest seriousness in the world of the tools they have equipped themselves to obtain water [...]. Recourse is made to salvaged objects, subjected to a metamorphosis, to an almost surrealist détournement with respect to their original function. Thus, shopping carts are used, a lot of rickshaws, strollers, wheelchairs, trabiccoli, and sometimes tools, machine platforms for transport or maintenance, everything that is on wheels and can function as a support for loads can become in the hands of the inhabitants a tool whose usefulness and effectiveness is praised and at the same time the toy that has the secret of transforming the most annoying and unpleasant task, the corvee of water, into a kind of game".

Sayad refers to an almost surrealist *détournement* probably because it is not the Situationist International but the Surrealist Movement that has established the most significant relationships with the contexts of decolonisation (think, for example, of the Césaire and the *Tropiques* magazine, of which Breton becomes a sort of spokesperson in Europe). But it is also true that perhaps the sense of what Sayad is showing can be traced back to the whole story of the historical avant-gardes and subcultures, from the dadaist *ready-made* to *do it yourself* punks. The car itself, for example, not the objects subjected to metamorphosis but the actual car, "has the *detournée* function of an instrument for the supply of water, because in the long run it costs less than the daily service that is paid to the children of the slum". And it is only by not grasping this *détournement*, therefore, by ignoring that the most equipped stalls in the internal market of the slum are those of salvaged objects and that the slum, in short, inherits from the colonial experience and decolonisation that immense "work on objects and on forms" with which Achille Mbembe wanted to represent the world described by Frantz Fanon in the first chapter of *The Wretched of the Earth* [10], that the car can be reduced to the mark of the scandalous tendency of these people to spend their money very badly, without giving priority to the expenses that would allow them to finally leave hell.

Incidentally, this is what still happens in the postcolonial dimension of the migrants' relationship with the smartphone, which once associated with the absence of a real condition of necessity, becoming the mark of a probable scam against transit or landing countries. And this while the web can only remain the diasporic medium *par excellence*, because only in cyberspace is it possible to shorten the distances that separate those who depart from those who remain and those who depart from one other, transforming devices such as satellite dishes or mobile phones into a "fragment of a house". The same fragment that allows the migrant to connect to the so-called *mobile commons*, that is to say the invisible and always up-to-date networks of information through which he can maintain contacts, plan travel and procure the jobs on which the success or failure of the journey depend [11].

In this light, beyond the most sensational and disturbing effects that pertain to the production and perception of decay, the making-bidonville itself can be interpreted as a practice of détournement through which the subject contributes to the formal determination of reality, without limiting himself to passively undergo its automatisms or the tendency to reproduce. If indeed the shanty town does not completely disappear but is "metamorphosed" into the new transit cities, as Sayad claims, it is not only in the form of a dispositif designed to make the subordination of the postcolonial condition chronic, but also as a resource for the survival of this same dispositif. Also *bidonvillisation* corresponds to a strategy of levelling out, in short, of transforming the space into a less hostile place and better corresponding to one's needs.

4 Perlocutionary Sequel

In conclusion, what Alfred Sohn-Rethel wrote about the city of Naples comes to mind: "No longer applied to the prescribed ends, the technique here undergoes the most unexpected deviations. As another example, the revolving motor released from the constraints of a destroyed motorcycle and delegated, by means of eccentric axles, to whip cream in a dairy could be useful. In this unexpected way, modern technology provides, despite itself, to this seventeenth century which unexpectedly continues, despite being equipped with electric trams and telephones, formidable aids and tools at the service of the freedoms of this life" [12]. In the specific case of the transit city, therefore, even *bidonvillisation* can perhaps be interpreted as an instrument "at the service of the freedoms of this life" or as a survival strategy for some of these freedoms on the borders of a formal citizenship that does not yet recognise them. But in order to grasp the dynamics of this relationship between concreteness and disavowal, and to imagine its developments, it does not seem necessary to me to resort to the ventriloquism of intentions or environmental causes, that is to say that it is possible to renounce the coordinates that John Austin uses to classify a speech act. Yet we continue to speak of performative citizenship in the sense of materially practiced and pregnant with consequences, even when the aim is not to retroactively act on the formal status of citizenship and is therefore not conceived as a political action.

Each statement, John Austin himself said, ends up exposing itself to the "production of a perlocutory sequel" which makes it temporarily operational even beyond the constitutive contexts, objectives and conventions [13]. For us, then, establishing that in the perlocutionary sequel of an action, political and legal transformations can also be determined without resorting to the self-referential expedient of ventriloquism, means recognising the right importance to the agency of a subject who has to do it despite everything.

References

1. Isin, E.F.: Performative citizenship. In: Shachar, A., Bauböck, R., Bloemraad, I., Vink, M. (eds) The Oxford Handbook of Citizenship. Oxford University Press, Oxford (2017)
2. Haraway, D.: The promises of monsters: reproductive politics for inappropriate/d others. In: Grossberg, L., Nelson, C., Treichler, P. (eds) Cultural Studies. Routledge, New York (1992)
3. Drissel, D.: Hip-Hop hybridity for a glocalized world: African and Muslim diasporic discourses in French Rap Music. Global Stud. J. **2**(3), 121–144 (2009)
4. Cohen, M., David, C.: Les cités de transit : le traitement urbain de la pauvreté à l'heure de la Décolonisation. Métropolitiques, 29 février (2012)
5. Tricart, J-P.: Genèse d'un dispositif d'assistance : les "cités de transit". Revue française de sociologie **18**(4) (1977)
6. Pétonnet, C.: Réflexions au sujet de la ville vue par en dessous. L'Année sociologique **21**, 151–185 (1970)
7. Guillaumin, C.: Racism, Sexism, Power and Ideology. Routledge, London and New York (1995)
8. Abdallah, M.H.: Cités de transit: en finir avec un provisoire qui dure. Plein droit **1**(68), 52–56 (2006)
9. Sayad, A.: Un Nanterre algérien, terre de bidonvilles. Editions Autrement, Paris (1995)
10. Mbembe, A.: Politiques de l'inimitié. La Découverte, Paris (2016)
11. Ascari, P. (ed.): Oggetti contesi. Le cose nella migrazione. Mimesis, Milan (2020)
12. Sohn-Rethel, A.: Das Ideal des Kaputten. Edition Bettina Wassmann (1995)
13. Austin, J., L.: How to Do Things With Words. Harvard University Press, Cambridge, MA (1975)

Pierpaolo Ascari teaches Aesthetics at the Department of Architecture, University of Bologna. He is author of *Ebola e le forme* (Rome, 2017), *Attraverso i confini. Letturastoria ed esperienza estetica in Stendhal e Flaubert* (Milan, 2018), *Corpi e recinti. Estetica ed economia politica del decoro* (Verona, 2019) and *The Adventure of Form: Aesthetics, Nature and Society* (Delft, 2021). He edited the volumes *Oggetti contesi. Le cose nella migrazione* (Milan, 2020), *Espressioni urbane. Muri sconciati, writing e street art* (Milan, 2022, whit Pietro Rivasi) and *I luoghi delle merci* (Milan, 2022).

Navigating Crises. Transient Communities for Urban Preparedness

Martina Massari

Abstract

The crisis has been challenging cities and urban planners for decades, producing theories, scenarios, and imaginaries aimed at governing its emergence and its consequences. The latest of such crises—the COVID-19 pandemic—seems to have once again brought attention to the unsolved urgency of planning to tackle the changes and stresses caused by insurgent events, which once again appears to involve cities and their spaces as holders of both the reasons and the possible solutions to the crisis consequences. Despite the abundance of possible approaches, however, the debate has not yet clearly highlighted the operative lessons learned from these challenges. In this vein, the urban studies debate has been reflecting on the possibility to act provisionally but in a preparedness perspective, opening alternative paths, rather than proposing solutions, planning for uncertainty and complexity with temporary means and actions. An attitude that requires both the flexibility and adaptability of consolidated urban systems and the affirmation and legitimisation of collective and practical instances in an operational institutional dimension. The article briefly reviews the relevant positions in the relationship between cities and crisis; next, it highlights the role, responsibilities and relevance of planning to inhabit the consequences of the crisis, from a preparedness perspective; lastly, it calls to consider the potential alternative answers resulting from interaction with temporary communities of practice.

Keywords

Preparedness · Transient communities · Adaptive planning · Urban crisis · Communities of practice

1 Crisis in the Urban World

"Crisis" comes from the Greek *krinō* meaning 'a decision made by a jury'. "Decision" has Latin roots, coming from *de-caedere*, to cut. In medical terms, this term is based on a practical concept, 'to sift', the activity of the Greeks when choosing grain seeds. In this definition, the crisis represents the moment of choice, of transition [1]. It indicates the unexpected modification of a physiological state, sudden and unpredictable because it radically changes previous conditions. In all aspects in which the word 'crisis' is conjugated, it indicates a distinct, sharp bifurcation, a change of conditions, as modification or

M. Massari (✉)
Department of Architecture, University of Bologna, Bologna, Italy
e-mail: m.massari@unibo.it

A. Borsari et al. (eds.), *TEMPORARY: Citizenship, Architecture and City*, The City Project 4,
https://doi.org/10.1007/978-3-031-36667-3_8

disappearance. The crisis age has been described as often characterised by acceleration, haste and focus on the present, [2] to the extent that—in some cases—the risk is to routinise the state of exception. At the same time, a countermovement of the opposite sign concerns the time of planning, design, decision-making, and proceduralisation, which is expanding as never before. We find ourselves employing more and more preparation and anticipation time, in contrast to the acceleration and presentism about which many agree is characterising times of crisis [3–5].

Against this, however, some authors have reflected on the slow pace of crises and on what causes this sharp bifurcation, focusing on the current critical situations that generate the vulnerabilities we are dealing with, as the result of a slow evolutionary path of our human species.

Baumann and Bordoni [1] express concern about the transformation of crises away from the form of conjunctures that appeared and disappeared in a limited period, evolving into a permanent state, which transforms by regenerating itself, influencing every sector of society in its evolution. Zardini [6] reflects on the path from today's crisis back to the 1970s, outlining what appears to be a long climatic-environmental, energy, and health crisis, which also affects institutions, professions and—as a consequence—the way we currently operate and design the living environment. Similarly, Serres [7] addresses the 2007 financial crisis. He explains that six major 'critical' events have been on the way for a long time and relate to the slow pace of the relationship between humanity, the Earth and technology:

- The largest population displacement in history (from the countryside to the city), contributing to provoking an ecological catastrophe that depends directly on anthropisation;
- The intensification and rapid pace of mobilisation, due to technological innovation in the transport sector;
- The development of health technologies, from which human bodies evolve into very different types from those of our ancestors;
- The greater demographic transition, exponential growth, and rapid ageing;
- The substitution of the collective, where the individual and the collective destinies were interconnected, with the connective, that allows relating at a distance and transforms ways and devices of production and reproduction of knowledge;
- The military and nuclear technologies that made humanity the most dangerous species.

As a general tendency, the occurrence of crises apparently did not introduce new ideas or paradigms but pushed toward policy change [8], favouring certain strategies and trends—which were already in place—until they became predominant. Following this thought, Latour [9] does not speak of a crisis as a contingent and removable event, but as a perennial condition. Other authors [1] observing the economic crisis of 2008 emphasise its deep roots and longer-term manifestations more than the temporary economic slowdown. This challenges the idea of crisis as a decisive moment of the upheaval of the political order, replaced by an understanding of crisis as a lasting, enduring [3, 10], condition from which there is no escape, and which requires us to navigate its uncertain outcomes.

These points convey the sense of the times in which we live, within a feeling of the unknown and uncertainty for the future, but at the same time a sense of urgency to act quickly [1]. The latest of such crises—the COVID-19 pandemic—seems to have once again brought attention to the unsolved urgency to tackle the changes and stresses caused by insurgent events, which once again involve cities and their spaces as holders of both the reasons [11] and the possible solutions [8] to the crisis consequences.

As expected, the consequences of COVID-19's most intense crisis years are superimposing themselves on the remnants of previous crises in the city spaces and social segments already hit hardest. At the same time, the insurgence of the pandemic crises has confirmed the pattern of what has been described as the bi-directional relationship between planning and crisis [8, 12].

It is widely acknowledged that urban planning is complicit in creating the conditions for the crisis which has deep roots in urbanisation [4], at the same time, the consequences of the crisis are exposing the weaknesses of planning and generating pressure for its re-thinking.

Following the first months of 2020 [5, 13], planning discourse has been re-shuffling well-established theories and models, which include the flexibility and adaptability of consolidated urban systems, the affirmation of "practical" and fast social answers and their legitimation in an operational institutional dimension. These positions are long-discussed challenges of urban studies, but they acquire new insights if reflected upon in a permanent crisis perspective.

The long-lasting temporality of the crisis urges on the one hand to re-think the temporaneous nature of the planning answers and on the other, to increase their territorial distribution.

Based on these premises, the paper reflects on two main issues:

- One is addressed by putting forward a "preparedness" dimension for urban planning, one that looks for a continuous capacity to respond appropriately to emergencies by making the uncertain future available for impermanent action in the present.
- The second issue is what interaction with communities of practice might offer concerning the re-connection with those socio-technical practices and solutions elaborated by actors directly engaged in the daily reproduction of responses to stresses and outbreaks. Communities of practice, as transient communities, are various forms of alliances built between territorial actors that are constituted and organised around shared issues.

These two postures will be explored to answer some guiding questions. How to develop the ongoing capacity to respond appropriately to uncertain futures, acting in the present? How to initiate exploratory conversations, augmenting the resilience of systems by increasing the number of agents ready to intervene in a crisis? And how to act in an experimental but permanent way? Looking at the crisis through the lens of urbanism can help answer to some of these questions, and lessons learned from experiments in our cities can indeed become sediments of future urban attitudes and approaches.

2 Preparedness as a Field of Experimentation for Planning

As has become increasingly clear in recent times, crisis events carry consequences of radical uncertainty that require new forms of rationality and a strong capacity for experimentation. Within this framework against which planning efforts must also be placed, falls the concept of *preparedness*. This keyword has been used to describe the adaptation and resistance to shocks, and crises, incorporating the concept of resilience and widespread adaptability. The concept was initially developed in relation to nuclear threats and bioterrorism, but was later extended to crisis management related to disasters, health threats and climate change risk [3]. The approach calls for moving from a state of emergency answer (typical of prevention and risk management) to vigilance and collective awareness for early detection of signals that lead to a system breakdown, and thus a crisis. Lakoff [14] distinguishes between prevention and preparedness as two ways of thinking about and intervening in the future of society. In prevention, a potential threat is considered a regularly occurring event whose probability can be calculated based on historical models and can be managed through risk distribution. Preparedness, on the other hand, applies to events whose regular occurrence cannot be mapped through statistical knowledge and whose probability cannot, therefore, be calculated. A threat is understood and managed as a certain event and, instead of limiting action in the face of uncertainty [13, 15], preparedness transforms potentially catastrophic threats into vulnerabilities that can be collectively mitigated [14]. Since the probability and severity of such events cannot be calculated, Lakoff argues that the only way to avoid disasters is to maintain a continuous capacity to respond appropriately by making an uncertain future available for action in the present. Similarly, Keck [16] uses the concept of preparedness to encourage a shift in focus from the short timeframe of emergencies to the long timeframe of ecologies [10]. This interpretation is useful to be explored as an orientation towards a specific way of understanding society and a specific form of evidence generation.

Anticipation of risk in the present [5] is an ecological notion [10] that carefully observes the sites where early warning signals (or sentinels) [16] are produced which, translated into an urban dimension, may represent the places that are most vulnerable in the event of emergencies or the most fragile populations that might be affected by their consequences. Keck, however, solicits that these signals, situated in local proximity to crisis events and consequences, are embedded into processes to assess alternative responses at different scales.

The concept of preparedness related to planning was once again introduced into the international debate immediately following the outbreak of the COVID-19 pandemic [3]. In a context of radical uncertainty and instability, and given the inability to plan today's actions and responses to cope with the ever-increasing environmental and social complexities, preparedness solicits thinking about the imponderability of crises that may arise, not aiming to avoid them but rather building a capacity to react that is valid in the most diverse situations.

Several authors [2, 17] call to adopt a preparedness perspective in planning. This directly relates to a resilient approach [17] that replaces the search for full predictive and preventive control of a situation, with the idea of dealing with both hidden development and sudden emergence with operational alternatives. However, planning in terms of non-linearity or uncertainty appears insufficient, as it would require constructing development opportunities which can continuously and iteratively reflect on themselves, improving the territorially distributed capabilities and competencies.

Pellizzoni and Balducci [3, 17] examine the transitional phases of disaster and crisis management: from the recognition of an external threat to be prevented (known); to the hidden (unknown) threats that emerge from the complexity of socio-ecological systems, for which one must prepare. Balducci further elaborates on this by stating that in today's society we cannot afford to be prepared for the 'known unknown', i.e. the risk we know will occur, but we must be prepared to face the 'unknown unknown', something unpredictable that should not find society ill-equipped, but ready to implement collective mitigation strategies. This, according to the authors, should inspire future policies and planning actions.

An interesting focus of this discourse is to see planning for preparedness as a form of shared responsibility while avoiding the risk of turning it into manipulative thinking [18]. This focus allows to seize the opportunities offered by the preparedness framework as a chance for planning to open up [19] to alternatives consistent with the permanent crisis situation. However, to seek adaptation to permanently open future risks generating more rigid and chrono-centric structures, or even formulating strategies that might confine innovation to predetermined future scenarios, reduces their flexibility and, consequently, their ability to rapidly adapt to unforeseen change.

Starting from the positions envisaged by preparedness, the risks and threads highlighted and its possible translation in a planning framework, the objective appears to become one of creating models capable of adapting, in a transformative and reversible manner, to something that has not yet been predicted and imagined, converting uncertainty and indeterminacy into an engine of innovation. As many suggest [17], not to renounce or eliminate the possibility of planning but to create complementary forms in terms of answers and capacities. Following this logic, preparedness requires being put to work, made operational and, above all, territorialised and contextualised [4, 20]. This resonates with the tendency to improve the territorially distributed capabilities and competencies [21] and to subvert the growing construction of tools to increase individual effectiveness [18] with a renewed emphasis on universal service provision and collective empowerment.

This direction calls for the need to broaden the base of agents ready (prepared) to provide alternatives in response to crisis events and long-term navigation. Henceforth, within the framework of preparation for inhabiting the crisis come reflections on the widespread approach to anticipation, with the construction of collaborations and alliances with urban actors who intervene in the city with their practices. This, besides being a well-established principle of urban planning, is part of the debate on socio-ecological resilience: this states that the adaptive capacity of a system increases as the number of actors able to collaborate with it increases and its ability to anticipate signals and consequently react to a crisis and its consequences increases.

3 The Construction of Impermanence: A Role for the Communities of Practice

As discussed above, the urban crisis with which the design professions are confronted is a perennial condition [9], offering opportunities for permanently temporary emergencies, but is no longer to be understood as a limited and reversible event. To plan for preparedness in the tense field of uncertainty, it appears necessary to expand the agency of citizens, a form of shared responsibility in the response to crises. Lessons learned from the ultimate urban crises [20] reinforce the hypothesis that an open framework for planning would be necessarily made operative by engaging with short-term, reversible, contingent, and dynamic solutions provided by the constellation of urban social actors. These solutions are often generated by informal organisations or even individuals self-activating and proposing alternatives to fill a gap left by the retraction of the public institution [22]. Through time, these experiences have been generating ad hoc, purpose-related, and transitional communities, forms of multi-agent collaborations that provide organisation to active citizenship, that are light, intentional, open, and reversible, and originated in specific space–time contexts. Drawing from the literature on an educational theory I will refer to them as communities of practice [23].

The term communities of practice is here employed to focus the attention on groups of subjects that, temporarily and intermittently, share values, needs and solutions for such needs [24]. It refers to ad hoc communities that exist to satisfy a particular problem and for a limited period [25], or transient communities [26] being activated by a project coalition in which the actors are involved in a common initiative, strongly sensitive to geographical proximity, but not exclusive to the local scale. They are called upon to give shape to spaces and devices, to reflect on the urban future or already ongoing transformations in the city through an innovative, sometimes radical perspective, investigating tactics that may not be codified, but are rooted in the urban population. This is the case, for instance, of energy communities, which are organised groups of citizens, homeowners or tenants who actively engage and cooperate to develop innovative forms of energy sharing [27]. Energy communities are configured as an aggregation of citizens for whom energy is a community asset, and it is widely recognised as one of the main leverages to act upon fighting climate change in the long run and achieving ecological transition. Energy communities can be interpreted as transient communities of practice for their purpose orientation that, through time, has generated new types of organisations. These are much more oriented toward value redistribution among participants, framed by specific types of governance, structures and purposes: cooperatives, foundations, limited partnerships, housing associations, non-profit customer-owned enterprises, public and private partnerships or public utility companies who "voluntarily accept certain rules for the purposes of shared common objectives" in relation to energy [28].

This indicates the coexistence of a group of individuals with like-minded interests in a communal space who interact not just in a defined geographical area, but mutually identify as a group which shares competencies born from common goals, values, and resources. Communities of practice are henceforth characterised by a social bond that is not stable, is both functional and solidaristic, and based on mutual recognition, exchange and sometimes gratuitousness. The latter is among the causes that leave these experiences to remain confined within contingent perspectives: as a matter of fact, they come into being in response to difficulties but do not always succeed in effectively carrying their impact towards a potential condition for change.

The community cooperatives (*cooperative di comunità*) in Italy, for example, are—still not-formalised—models for the aggregation of localised enterprises, associations and institutions, organised to be producers and users of goods or services, fostering synergy and opportunities for growth and cohesion. They are becoming a model for their capacity of responding to crises thanks to their ability to internalise emerging needs by finding a solidarity-based model of response to them. According to Wenger, members of communities of practice structure their identity from their experiences in the community and the meaning they attach to their actions within their context. It can be said that communities of practice are able to produce an alternative vision of (temporary) citizenship, one based on the defence, production and management of collective interests and goods.

The most recent crisis has accelerated and, in some ways, prototyped new networks and coalitions for communities of practice, oriented to expand their role and influence, encompassing heterogeneity, capable of sharing common missions. Alliances that involve different actors who share both means and common goals. Many of them provide services and products, involving experiences that include communication activities, set-ups, and collaborative practices implemented thanks to the commitment and voluntary work of citizens and social groups.

Among these new prototypes we can mention the experience of Ethical Delivery (*consegne etiche*), an alliance between workers, shop owners, and restaurateurs in Bologna (Italy), to respond to two challenges: a contingent one, to help the delivery of goods in the short term during the COVID-19 pandemic (see Fig. 1); and a long-term one, to reduce the prevalence of large platforms which have long been deficient in terms of workers' rights. On a similar note, see the case of the South Working Association [29] which promotes innovative formulas of business development, self-regulation and management of spaces and services for de-localised smart workers who, during the hardest months of the COVID-19 lockdown, were forced to work from their hometown (often located in remote places from where they had moved). The

Fig. 1 Consegne Etiche—Margherita Caprilli for Fondazione Innovazione Urbana

experimentation has been so successful that it the pilot network has become a stable service for workers re-locating to work remotely in their (southern) hometowns (see Fig. 2).

The actions of ad hoc, transitional communities of practice take shape through interaction. They are thus originated by the actions of the people that share an issue and the intention to address it collectively, generating joint meanings, albeit pro-tempore, in a non-stable form. This learning outcome should not be seen as an interference in planning but at the same time, communities of practice should not be considered as a homogeneous "whole" responding "mechanically" and automatically to a systemic need or to unmet crisis consequences. Indeed, the challenge is to formulate strategies that do not confine them to predetermined future scenarios, as this would risk reducing their flexibility and, consequently, their ability to adapt quickly to unforeseen circumstances, acting also as signals for the consequences of an imminent crisis.

4 As an Open Conclusion, Can Communities of Practice Operationalise Preparedness in Planning?

Western cities have long been going through a field of tensions due to a series of pressures and stresses generated by various crises (economic, financial, global, climate and, most recently, health). The consequences have deposited in the living environment a series of challenges that have tested mainstream planning and decision-making processes, but also risk management practices: declining public funding in the provision of services, the increased prominence of non-traditional urban actors in the production of welfare, the slowdown in the housing market, political instability, and distrust in policies.

While navigating within the ultimate crisis, we wonder what to learn, what to leave behind, but above all how to adapt to its permanent consequences with which we have to live.

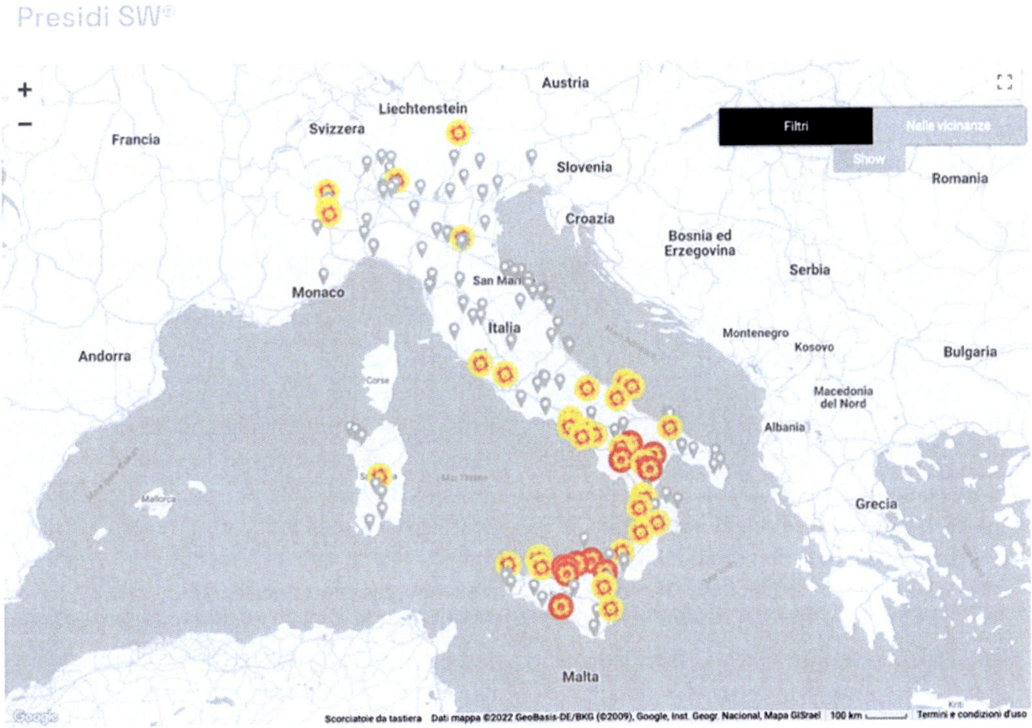

Fig. 2 Southworking online map of community outposts. https://www.southworking.org/mappa-dei-presidi/

Planning for preparedness exhorts territorial planners, designers, and policy makers to "stay" in the crisis and its consequences. Broadening the scope of action and awareness of responsible and vigilant actors regarding crisis signals could go in this direction. In the background, a number of issues emerge from the lessons learnt from transient communities of practice, from which to operationalise a preparedness approach in planning, design and policy making.

The first aspect to be taken into account in the actions of communities of practice is temporariness. The inherent logic of constituting themselves (formally or otherwise) as a result of a need that may be contingent, also characterises their results: their actions are more tactical than strategic, pro-tempore rather than permanent. At a time of transformation of planning, design and policy making, more oriented towards strategic than spatialised, normative arrangements, the temporariness of practices and the possibility of building alliances with them, deserve attention and discussion.

The second aspect concerns their geography. They are both strongly spatialised and localised experiences (such as community cooperatives, which refer to a defined territorial context), but also collaborative networks that emerge in multiple contexts at the same time (the South Working example is paradigmatic of the ability of a localised movement to act at the same time in strong connection with the national dimension, to the point of touching on global issues). This is a geographical dimension led by proximity, where the term does not only indicate spatial closeness, but a communion of intentions and goals, thanks to daily (or sporadic) contact and exchange. The increasing degree of complexity and the critical mass determined by the practices of transient communities suggest the adoption of approaches able to deal with new spatialities beyond administrative borders and oriented to build a more coherent spatial logic, responsive to the real and effective living territory.

The experimental approach is the third point of discussion. This is a way of working that allows practices to act on a small scale, through prototypes to be tested and then upscaled. The case of Consegne Etiche in Bologna followed a similar path, first testing and prototyping the possibility of addressing the ethical issues of the deliverymen's work by going through a process of assessment and constant evaluation, first on a small scale and then at city level. Experimentation allows to adjust, think laterally, try new possibilities and fail, in order to expose other realities with alternative points of view. This approach might mirror an inquiry process for planning, which is essential to exercise a real capacity to review planning, design choices and actions. Policy understood as a process of enquiry, however, is necessary to activate dialogue and transactions between the many actors involved in urban reality. This involves sharing a problematic and strategic approach that clarifies the issues at stake and solutions through provisional projects, reflecting an inductive research methodology.

The last consideration concerns the governance of communities of practice, as a hybrid, multi-actor contamination between public, private and social actors. The governance model envisages that the parties interested in collaborating take on a characteristic (or a part) of the actors with whom they collaborate: a hybrid set-up that favours institutional change by affecting —albeit temporarily—the balance of power relations in planning processes.

These preliminary points of reflection allow to consider the equipment and skills of the ordinary everyday practices of communities to be repurposed in response to the extraordinary nature of the crisis, in a planning setting, with an argumentative attitude, orienting rather than determining, selecting what is indispensable, but also saying what is provisional and therefore treated differently. Furthermore, it is a useful overview for understanding how to pick up the thread of a great wave of public planning which, at least in Europe, will be able to benefit from investments and resources on a considerable scale.

Interpreting how communities use (and struggle for) daily and temporary (even precarious) solutions to crises in an organised and engaging way can reveal several insights. On the one hand, the needs emerging from crisis situations are often interconnected not only at a neighbourhood scale but follow broader geographies. On the other hand, unexpected relationships and interactions emerge between the built environment and the way people inhabit it in response to changed living (and working) needs. The way transient communities of practice operate can ultimately contribute to both the theory and practice of the co-production of cities and territories.

In conclusion, a research perspective is identified, whereby transitional communities of practice may become the privileged observatory where attitudes, but also alternatives and even inspirations to crisis preparedness policies, can be recognised. Practices that could ultimately inspire a broader change, especially in the delicate management of risks: from a controlling attitude towards an approach centred on institutional adaptation, to coexistence and governance for navigating the urban and territorial crisis.

References

1. Bauman, Z., Bordoni, C.: State of Crisis. Wiley (2014)
2. Pasqui, G.: Coping with the Pandemic in Fragile Cities. Springer, Milan (2022)
3. Pellizzoni, L.: The time of emergency. On the governmental logic of preparedness. AIS **16**, 39–56 (2020). https://doi.org/10.1485/2281-2652-202016-3
4. Connolly, C., Ali, S.H., Keil, R.: On the relationships between COVID-19 and extended urbanization. Dialogues Hum. Geogr. **10**(2), 213–216 (2020). https://doi.org/10.1177/2043820620934209
5. Dodds, K., et al.: The COVID-19 pandemic: territorial, political and governance dimensions of the crisis. Territory, Politics, Governance **8**(3), 289–298 (2020). https://doi.org/10.1080/21622671.2020.1771022
6. Doglio, F., Zardini, M.: Dopo le crisi. 1973, 2001, 2008, 2020, vol. 1. LetteraVentidue Edizioni (2021)
7. Serres, M.: Times of Crisis: What the Financial Crisis Revealed and How to Reinvent Our Lives and Future. Bloomsbury Publishing USA (2013)
8. Ponzini, D.: Introduction: crisis and renewal of contemporary urban planning. Eur. Plan. Stud. **24** (7), 1237–1245 (2016). https://doi.org/10.1080/09654313.2016.1168782
9. Latour, B.: Facing Gaia: Eight Lectures on the New Climatic Regime. Wiley (2017)
10. Bifulco, L., Centemeri, L., Mozzana, C.: For preparedness as transformation. Sociologica, 5–24 (2022). https://doi.org/10.6092/ISSN.1971-8853/13939
11. Harvey, D.: The urban roots of financial crises: reclaiming the city for anti-capitalist struggle. Socialist Register **48** (2012)
12. Tulumello, S., Saija, L., Inch, A.: Planning amid crisis and austerity: in, against and beyond the contemporary conjuncture. Int. Plan. Stud. **25**(1), 1–8 (2020). https://doi.org/10.1080/13563475.2019.1704404
13. Ibert, O., Baumgart, S., Siedentop, S., Weith, T.: Planning in the face of extraordinary uncertainty: lessons from the COVID-19 pandemic. Plan. Pract. Res., 1–12 (2021). https://doi.org/10.1080/02697459.2021.1991124
14. Lakoff, A.: Unprepared: Global Health in a Time of Emergency. University of California Press (2017)
15. Moroni, S., Chiffi, D.: Complexity and uncertainty: implications for urban planning. In: Portugali, J. (ed.) Handbook on Cities and Complexity, pp. 317–330. Edward Elgar Publishing, Cheltenham Glos (2021)
16. Keck, F.: Avian Reservoirs. Duke University Press (2020)
17. Balducci, A.: Planning for resilience. In: Risk and Resilience, pp. 15–25. Springer (2020)
18. Amin, A.: Surviving the turbulent future. Environ. Plan. D **31**(1), 140–156 (2013). https://doi.org/10.1068/d23011
19. Sennett, R.: Costruire e abitare: Etica per la città. Feltrinelli Editore (2018). [Online]. Available: https://books.google.it/books?id=l9HVDwAAQBAJ
20. Acuto, M., et al.: Seeing COVID-19 through an urban lens. Nat. Sustain. **3**(12), 977–978 (2020). https://doi.org/10.1038/s41893-020-00620-3
21. Coaffee, J., Lee, P.: Urban Resilience: Planning for Risk, Crisis and Uncertainty. Macmillan International Higher Education (2016)
22. Moulaert, F., MacCallum, D., Hillier, J.: Social innovation: intuition, precept, concept. In: The International Handbook on Social Innovation: Collective Action, Social Learning and Transdisciplinary Research, vol. 13, pp. 13–23 (2013)
23. Wenger, E.: Communities of practice: learning as a social system. Syst. Thinker **9**(5), 2–3 (1998)
24. Pasqui, G.: La città, i saperi, le pratiche. Donzelli Editore (2018)
25. Sloep, P., et al.: Ad hoc transient communities to enhance social interaction and spread tutor responsibilities, p. 7 (2007)
26. Kester, L., Berlanga, A.J., Sloep, P.B., Brouns, F., van Rosmalen, P., Koper, R.: Ad hoc transient communities: towards fostering knowledge sharing in learning networks, p. 17 (2008)
27. Walker, G., Devine-Wright, P.: Community renewable energy: what should it mean? Energy Policy **36** (2), 497–500 (2008)

28. Moroni, S., Alberti, V., Antoniucci, V., Bisello, A.: Energy communities in the transition to a low-carbon future: a taxonomical approach and some policy dilemmas. J. Environ. Manage. **236**, 45–53 (2019). https://doi.org/10.1016/j.jenvman.2019.01.095
29. Associazione South Working: South Working: Per un futuro sostenibile del lavoro agile in Italia. Donzelli Editore, Roma (2022)

Martina Massari, architect and Ph.D. in Urban Planning. Currently Research Fellow at the architecture Department of the University of Bologna and freelance architect/planner. She has been a Research Fellow at the Chair for Regional Building and Urban Planning at the Leibniz University of Hannover. She is part of the "CPCL Agency", the "TRACE Team" and the "Collaborative and Adaptive Cities" research groups of the University of Bologna.

Energy Community and Citizenship as Enabling Actions for Integrated Energy Plan Implementation and Urban Energy Transition

Serena Pagliula

Abstract

One of the main challenges of the future is to create climate-neutral and smart cities—with zero CO_2 emissions by 2050. Cities have to be conceived, planned and designed in order to remain competitive and survive growing populations, scarce resources and changeable and unpredictable built environments by putting sustainability, health and the quality of citizens' lives at the centre. Starting from the climate-neutral city vision, this paper aims to investigate the link between integrated energy plans developed by cities and the emergence of energy community as enabling action for the implementation of plans. Methodological examples of integrated energy plans, such as the SmartEnCity (SmartEnCity—Towards Smart Zero CO_2 Cities across Europe, H2020 GA691883) methodology and the Sustainable Energy and Climate Action Plan (SECAP) developed by the European Covenant of Mayors for Climate and Energy, are considered. The analysed GECO (Green Energy Communities, EIT Climate KIC TC_2.2.15_190736_P125-1) and GRETA (Green Energy Transition Actions, H2020 GA101022317) research projects attempt to solve the complex issue of coordinating multiple stakeholders, different sources of funding and innovative technologies under the umbrella of a legislative and contractual framework in a replicable scheme all over Europe. The lesson learnt of this paper contributes to highlighting how integrated energy plans can favour the emergence of energy community, as legal entity, and energy citizenship, as new definition of consumer, and help to move from the planning to implementation phases, as well as opening to new questions that will be developed in future research.

Keywords

Climate-neutral city · Integrated energy plan · SECAP · Energy community · Energy citizenship

1 Introduction

Cities play a crucial role in the energy transition process, considering that 75% of European citizens live in urban areas, a percentage that is expected to increase to 84% by 2050 [1]. For a city to be defined as climate neutral it is necessary that: (1) Its carbon footprint and energy demand are minimised using innovative technologies that reduce energy demand and promote

S. Pagliula (✉)
Department of Architecture, University of Bologna, Bologna, Italy
e-mail: serena.pagliula2@unibo.it

A. Borsari et al. (eds.), *TEMPORARY: Citizenship, Architecture and City*, The City Project 4,
https://doi.org/10.1007/978-3-031-36667-3_9

citizen awareness; (2) The supply of energy is entirely renewable and clean; (3) Local energy resources are intelligently managed by conscious citizens as well as by private and public stakeholders [2].

According to the European Green Deal, cities are called to achieve climate neutrality by 2050. In line with the vision of the *EU Mission for 100 climate-neutral cities by 2030—by and for the citizens*, this ambitious goal cannot be achieved without an ecological and cultural change involving all interested stakeholders, including public administrations, businesses, universities and citizens [2]. At the Conversations on the New European Bauhaus conference on 22 April 2021, Ursula von der Leyen stressed that the transition towards a climate-neutral Europe is synonymous with cultural transformation, since the climate-neutral target must be firstly understood, felt and shared with people in order to be achieved [3].

Public administrations primarily play a driving role in the planning, design and implementation of sustainable actions aimed at energy efficiency, optimisation of available resources and migration to renewable energy production sources in compliance with current legislation. The development of integrated energy plans is considered an effective strategy adopted by municipalities to coordinate medium-long term intervention actions, outlining how cities can reach their energy commitments by 2030 and 2050. The integrated energy plan and its implementation is indeed defined by the *Smart Cities Marketplace* as an "*iterative sequence of phases from planning to implementation and replication*" [4] (p. 2), which includes in each phase a long-term perspective as well as interested multiply policy levels, combination of sectors such as energy, transport and ICT, involved stakeholders and financial aspects.

The planned actions include measures for citizen involvement based on the creation of an environment of trust by public administrations, in order to make citizens themselves aware and responsible for the implementation of energy efficiency actions, while at the same time adopting sustainable models of behaviour. Citizen involvement can be supported by the creation of energy communities. The energy community is a legal entity which, in accordance with the European and national laws in force [5], is based on open and voluntary participation, is authorised to independently produce, consume and sell energy from renewable sources, and has the aim of providing environmental, economic and social benefits to the community, for its stakeholders, its members or the local areas in which it operates. The birth of an energy community involves the aggregation of a certain number of prosumers willing to share plants for the production of electricity produced from renewable sources [6].

The concept of energy community is strongly connected to the energy citizenship one, which represents a new definition of consumers [7], and aims to further explore the sociological and cultural as well as technological aspects of shared energy. The research under development in the H2020 GRETA project—Green Energy Transition Actions (GA101022317)—identifies the structural and dynamic aspects of the emergency of energy citizenship at different geographical levels, analysing relevant factors such as the presence of planning, climate, demography, technologies for energy efficiency and renewable energy production, morphology of the built environment, funded projects and initiatives. In particular, the Italian study is applied to the Roveri-Pilastro urban area, a mixed-use district in the northeast of Bologna.

2 European Energy Directives and Initiatives

This section provides an overview of the energy directives and initiatives to be considered to frame the climate-neutral and smart city vision. The European Green Deal, announced in 2019 in a European Commission (EC) communication [8], is one of six priorities for the years 2019–2024 and defines a set of political initiatives by the EC to achieve climate neutrality in Europe by 2050. According to the European Green Deal vision, Europe has to be transformed into a modern, resource-efficient and competitive economy, characterised by zero emissions of greenhouse gases, economic growth not linked to resource use, and people inclusion, in line with the commitments of the Paris Agreement.

As a follow up to the European Green Deal and to implement the increased ambitions, in July 2021 the EC adopted a package of proposals in a review of the entire EU *2030 climate and energy framework* [9], explained in the *2030 Climate Target Plan* [10] and transposed into the first *European Climate Law* [11]. The adopted proposals include legislation on effort sharing, land use and forestry, renewable energy, energy efficiency, emission standards for new cars and vans, and the Energy Taxation Directive. In particular, the scope of the so called *Fit for 55 package* is to adapt European climate, energy, land use, transport and taxation policies to reduce net greenhouse gas emissions by at least 55% by 2030, compared to 1990 levels [12].

The climate-neutral and smart city concept strongly emerges within the European Mission *100 Climate-neutral Cities by 2030—by and for the Citizens*, launched in September 2020 [2], and has been better defined and concretely explained in its path within the *Info Kit for Cities* [13]. Hanna Gronkiewicz-Waltz, Chair of the Mission Board for Climate Neutral and Smart Cities, addressed cities, saying: "*We expect you to implement this Mission with the same determination as the Americans did with their Moonshot. The climate-minded transformation of cities goes far beyond the idea of the Man on the Moon. This is The Mission of our times!*" [2]. The aim of this mission is to select, through a call for expression of interest closed in February 2022, 100 climate neutral cities able to reach climate neutrality by 2030, twenty years before the deadline set by the European Green Deal. This mission contributes to speed up the energy transition process in cities considering the worsening environmental crisis, the change of habits and health issues following the pandemic emergency, and the shock of the Ukrainian conflict that makes it fundamental for Europe to become independent from Russian gas, as explained in the *REPowerEU Plan* published in May 2022 [14].

The eligibility criteria to join the *100 Climate-neutral and smart cities by 2030* call were two: (1) City of at least 50,000 inhabitants in order to maximise the impact of the energy measures and, (2) The ambition to become climate neutral demonstrated through current levels of GHG emissions, the level of preparedness with existing plans and policies, targets, governance structures and administrative capacity, and the efforts to involve citizens and other stakeholders in both planning and implementing phases [2]. The intention of the EC is to show with real case studies of more advanced cities that urban energy transition is possible as well as being mandatory and track the path for other European cities until 2050. What appears fundamental and, at the same time critical, to become a climate neutral city is the full awareness of current GHG emissions, a forecasting and planning approach to manage and implement energy implementation actions, and the citizenship involvement process. In line with the Green Deal objectives, many other initiatives have emerged for supporting city journeys towards carbon neutrality. One of the most known is the *Covenant of Mayors*, an alliance that offers guidance and methods for developing Sustainable Energy and Climate Action Plans (SECAPs). Other initiatives include the *Smart Cities Marketplace* that creates networks between smart city stakeholders and investors; the *Renovation Wave* that aims at greening buildings; the *New European Bauhaus* that connects the *Green Deal* to daily lives and living spaces; and the *JPI Urban Europe* knowledge hub for urban transitions.

3 Methodological Strategies and Application of Integrated Energy Planning

The integrated energy plan is considered as a methodological strategy in the hands of municipalities to trace the energy transition path of cities, enabling them to dialogue with the surrounding territory. To be effective and achieve energy targets, the plan has to be integrated into governance activities and co-designed with local stakeholders, on the basis of a bottom-up approach which allows, in addition to directly detecting energy input data, sharing intervention actions with the territory. Furthermore, it is important that the energy plan is connected with other urban planning tools, such as general urban plans, mobility plans and building regulations. A fundamental aspect not to be overlooked in the planning process is the involvement of citizens, associated with an information and awareness-raising action that starts from the public administration and invests the private sector, thus encouraging new investments and projects [15, 16].

It is necessary to create a direct channel of communication with citizens from the beginning of the planning process, for example through associations. Moreover, it is important to include them in the decision-making process, considering that many of the actions that involve energy consumption are related to their modus operandi in everyday life and that various benefits in terms of energy efficiency are intended for them. The ICT tools that have spread quickly, also due to social distancing as a result of the COVID-19 crisis, are reflected in the increased dissemination of information and usability of data, in support of a conscious and continuous commitment of citizens in the energy transition process. Through the planning activities carried out and the planned intervention actions, including the creation of an energy community, it is hoped to successfully achieve the energy objectives of the Green Deal and transform environmental challenges into opportunities in all sectors and policy areas, making the energy transition fair and inclusive for all.

3.1 The Covenant of Mayors Programme and SECAPs

In October 2015 the new integrated Covenant of Mayors for Climate & Energy [17] was presented, born from the union of the Covenant of Mayors and Mayors Adapt, with energy objectives within the *2030 climate and energy* [10] and *Fit for 55* [12] legislative framework.

The Sustainable Energy and Climate Action Plan (SECAP) represents one of the urban energy planning strategies made available by the European Commission for public administrations. Based on Covenant of Mayors (CoM) requirements, local authorities that have developed a SECAP must report their status of implementation to the CoM every two years. In Europe, the signatory administrations of the CoM are over 10,000 and over 6000 of these have presented action plans [18]. In Italy, the percentage of signatory municipalities is 58% [19]. These data are indicative of municipalities' ambition to invest in energy planning. According to recent updates from the CoM, the percentage of Italian signatories that have submitted action plans is around 81%, while the percentage of Italian signatories that have produced monitoring reports is around 21% [20]. These data represent a rather low progress of the monitoring phase, indicative of the fact that plans have not always translated into numbers, and actions, when realised, have not been registered. Furthermore, the developed strategic vision of the SECAP scope can be enhanced using innovative elements of strategic energy planning with long-term decarbonisation goals and including holistic energy systems and improved energy scenarios [18].

Thanks to the research developed in the context of the European Horizon 2020 projects, the cities involved have been the subject of experimentation and application of advanced integrated energy plans, such as the case of Lecce in the SmartEnCity project.[1] The Lecce SmartEnCity Integrated Energy Plan (IEP) is used as an example of performed energy planning activity. Another example of an already implemented SECAP taken into consideration is that of Bologna Municipality [21], approved in April 2021.

In both plans (SmartEnCity IEP for Lecce and Bologna SECAP), energy community has been included as a strategic action to increment the use of renewable energy and allow implementation of the citizen engagement process. In the

[1] https://smartencity.eu/about/follower-cities/lecce-italy/, last accessed 2022/06/27.

Bologna SECAP, it is specified that the energy community creation is funded and supported by the GECO project (Green Energy Communities, EIT Climate KIC TC_2.2.15_190736_P125-1) with the objective to create the first case study of energy community in the Pilastro-Roveri area of Bologna [21, 22].

3.2 The SmartEnCity Methodology and Experience

The SmartEnCity project aims to develop a systemic, adaptable and replicable approach to transform European cities into sustainable, intelligent and energy efficient environments through urban planning integrated with the optimisation of urban energy consumption and the production of renewable energy.

The research project is based on developing and implementing the Smart Zero CO_2 concept in the three lighthouse cities of Vitoria-Gasteiz (Spain), Tartu (Estonia) and Sonderborg (Denmark) and replicating the energy planning approach in the two follower cities of Lecce (Italy) and Asenovgrad (Bulgaria). The Smart Zero CO_2 cities have the following characteristics: carbon footprint brought to zero; energy demand brought to a minimum level using control technologies that can save energy and increase awareness; completely renewable and clean energy supply; and energy resources efficiently managed by citizens, public and private stakeholders. The main object of the experimentation is to include energy demand, supply and management, apply innovative technologies, and integrate participatory processes is the cities' energy systems. The research has been conducted together with continuous feedback from the real case studies. According to the deliverables published within the Work Package 8 "*Replication to Followers and Smart Cities Network*",[2] what impresses most about the SmartEnCity methodology is that the city system is considered an aggregate of components, all contributing to energy transition and decarbonisation, with the aim to get municipal employees, planners and technicians to consider the decarbonisation process from an integrated approach and involve all municipal sectors (i.e. energy, mobility, ICT, governance, etc.). This methodology, as output of the project, provides a guide for cities in their ambitious urban transformation and a replication toolkit, called *SmartEnKIT*,[3] made available to all cities.

The benefits of an integrated urban planning approach for cities can be summarised in the following points: (1) Development of cross-sectoral goals and monitoring systems; (2) Implementation of projects that involve the knowledge and perspectives of different disciplines and public and private stakeholders; (3) Support to use the available budgets and capacities in an efficient way to implement city goals by joining capacities and funds and by reducing barriers among municipal sectors.

3.2.1 Lecce Municipality Case Study

The city of Lecce is a partner and follower city of the SmartEnCity project. Lecce was involved in the integrated energy planning activities on the basis of the approach already tested on the lighthouse cities Vitoria-Gasteiz in Spain, Tartu in Estonia and Sonderborg in Denmark. Lecce's SmartEnCity Integrated Energy Plan (IEP) is based on a twofold objective: on the one hand, the definition of specific strategic lines aimed at reducing CO_2 emissions, and on the other, the coordination of municipal offices and public and private stakeholders active in the area, in order to optimise the positive effects of the energy transition. The methodological phases for the plan development are the following (Fig. 1):

- Phase 1—"Engage": Lecce created a technical team dedicated to the project for managing energy planning and monitoring activities.
- Phase 2—"Analysis": The technical team collected all the input data and information needed for defining the city's characteristics, such as:

[2] https://smartencity.eu/outcomes/public-papers/, last accessed 2022/06/27.

[3] https://smartencity.eu/outcomes/smartenkit/, last accessed 2022/06/27.

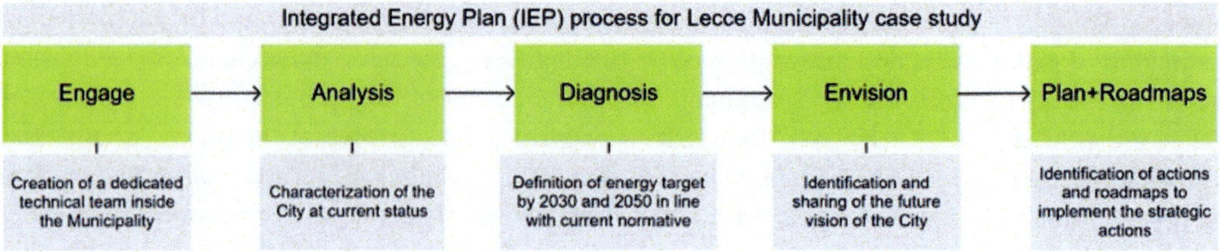

Fig. 1 Integrated energy plan process for Lecce

- Analysis of plans from other cities already implemented in Europe;
- Direct sources from municipal offices;
- Sustainable Energy Action Plan—SEAP with 2020 energy target;
- Data from mobility plan;
- Municipal building regulations;
- Energy balances;
- Results from the H2020 PLANHEAT project (GA723757) from simulation performed on Lecce GIS file integrating alternative energy supply solutions produced by advanced technologies for heating and cooling, and balancing of forecasted energy demand.

• Phase 3—"Diagnosis": Lecce's strategic question was: *How can we reduce CO_2 emissions by 40% up to 2030?* This target was in line with the Integrated National Plan for Energy and Climate objectives [23], with reference to the energy directives in force when the plan was developed.

• Phase 4—"Envision": the definition of the future strategic vision was developed with the support of local stakeholders, invited through an expression of interest to participate in a workshop, using a SWOT analysis as a comparison methodology to identify the strengths, weaknesses, opportunities and threats in the sectors of energy, mobility, ICT and governance, alongside the identification of trends in smart cities. The stakeholders that replied to the expression of interest were selected by the technical team considering expertise and need to cover the strategic areas of the plan: energy; urban lighting; mobility; ICT and new technologies; natural resources; waste management; government; and people.

• Phase 5—"Plan": the energy plan developed by the technical team integrated the activities carried out in the previous phases, considering the results achieved by the other consortium cities and resulting from the workshop organised with the local stakeholders.

The Lecce SmartEnCity IEP includes a dedicated part called *Roadmaps* that represents the final step of the planning process, finalised in January 2021. On the basis of the inputs collected by the local stakeholders involved, the Lecce technical team selected and further expanded four strategic implementation actions that were fundamental to reach the energy target. Some of them are ongoing (i.e. energy efficiency improvement of public real estate), others need to be further exploited (i.e. production of energy from renewables, promotion of internal training courses for the specialisation of municipal employees in smart city topics) and others are yet to start (i.e. creation of an energy community towards the development of a positive energy district). The creation of the energy community represents the strategy to reply to different needs and resolve complex issues such as citizen engagement and the application of renewable technologies at district level. The *Roadmaps* section of the plan provides specific details on municipal sector responsible of the action, possibly involved stakeholders; approach for stakeholder engagement; current status of the action; draft of project timeline; steps for the action's implementation; input of technologies that can be used and available funding.

4 Roadmap Towards Energy Community and Citizenship

The need to intensify citizen involvement in planning and implementation activities appears evident from the monitoring of greenhouse gas emissions within the SECAPs. Palermo et al. [24] analyses 315 signatories of the Covenant of Mayor and the result of this research shows that the main focus of policies and efforts is the buildings sector. Moreover, from this analysis it is interesting to see that municipal buildings is the field that includes the majority of policy tools' application, comparing to residential and tertiary buildings. This result is coherent with the highly developed residential building stock which is mainly responsible for CO_2 emissions in European cities [24]. If we also consider the case of Bologna Municipality, according to the data reported on the Covenant of Mayors webpage,[4] the majority of greenhouse gas emissions come from the residential sector. These data demonstrate that actions performed by municipalities and public entities to involve citizens and private sector operators and implement energy efficiency projects for private buildings, need to be further enhanced.

In this scenario, the *Clean Energy for all Europeans package* [5], the *Renewable Energy Directive (Directive (EU) 2018/2001)* and the *Common rules for the internal market for electricity (Directive (EU) 2019/944)* launched the concept of the energy community, where citizens assume the active role of prosumers to produce and share renewable energy. The energy citizenship definition, as researched within the ongoing GRETA project, strongly marks the need to raise awareness and involve citizens as crucial partners in the urban transformation process [25, 26].

While Lecce Municipality has only planned the energy community as an implementation action, Bologna Municipality has already adopted the new SECAP including the energy community as an intervention action, thanks to funding obtained within the GECO and GRETA projects. As shown in these two research projects, the energy community is strictly linked to energy citizenship creation, finding their experimentation in the common Italian case study represented by the Pilastro-Roveri area of Bologna Municipality. The analysis performed on the case study of the Pilastro-Roveri area aims to investigate the main characteristics of energy citizenship, identify problems, frame solutions and reach consensus on the roadmaps for change, formalised through the development of Energy Citizenship Contracts [26]. The energy communities in which the active participation of citizens takes place, who identify and converge in the establishment of energy citizenships regulated by specific contracts, become a fundamental and effective action for contributing to the increase of the flexibility of the electricity system and consequently the number of local prosumers. The Bologna SECAP and the implementation actions have been effective in getting the city selected to become one of the 100 climate-neutral and smart cities by 2030, providing an example of good practice to other cities.

5 Lesson Learnt and Conclusion

Energy transition in cities needs to be guided in technical, economic and financial aspects through an integrated energy plan including implementation actions. Supporting replication of good practices in cities is necessary, because the case studies provide a concrete example that can be better understood by other cities.

An integrated energy plan is not just a report, it can be considered a live process that starts from a change of mentality among both political and technical actors within the municipality, as experimented within the SmartEnCity project. The change consists in inverting the way of acting, no longer starting from a single resolution of emergencies or a rush to already available finance, but foreseeing future scenarios in a critical way, and adapting and moving towards a shared vision of the city based on new principles, on the basis of an accurate analysis of energy

[4] https://www.covenantofmayors.eu/about/covenant-community/signatories/action-plan.html?scity_id=11963, last accessed 2022/06/27.

baselines. The development of the shared vision is important to imagine a new improved city in 10, 20 and 30 years. The activation of this process of change can start by defining capacity building pathways for decision makers and municipal employees. In the case of Lecce Municipality, it was very impressive to go and see the implementation actions in Victoria and Tartu aimed at stimulating municipal representatives to think about a new innovative vision of the future city. The next necessary effort is intercepting funding and knowing how to channel it into projects, following and updating the integrated energy plan.

Standardised tools to automatically monitor indicators, such as energy consumption, CO_2 emissions and renewable energy production by sector, can be very helpful to municipalities to constantly control and update the data included in the energy plan and to provide information on which to base future actions. The monitored data can measure and explain the plan in terms of progress or regress and improve the roadmap if necessary.

This paper concludes with some issues that need to be further investigated in future research and faced by and with municipalities. Integrated energy plans can be further improved using a more concrete approach, linked to a data driven decision-making progress. In this sense, data monitoring and tools can assume a more significant and important role to support the responsible actors of municipalities in collecting and monitoring energy and climate data, and updating plans. An analysis of the most used/user-friendly tools can help the data collection exercise. Furthermore, it will be interesting to distinguish the outcomes that can be effectively relevant for municipal technical teams, and the outcomes that guide governance.

Energy community and citizenship, in their legal, technical and sociological dimensions, can become the framework both for collecting energy consumption and production data, and for engaging citizens as well as municipalities to directly detect energy data from private/public buildings. Under the umbrella of the GRETA project, the connection between energy data collection and monitoring, and energy community and citizenship creation, is investigating by evaluating the possibility to include and formalise energy monitoring activities within Energy Citizenship Contracts.

References

1. United Nations Department of Economic and Social Affairs, Population Dynamics—World Urbanization Prospects (2018). https://population.un.org/wup/Download/. Last accessed 27 June 2022
2. European Commission: 100 Climate-neutral Cities by 2030—by and for the Citizens. https://ec.europa.eu/info/publications/100-climate-neutral-cities-2030-and-citizens_en. Last accessed 27 June 2022
3. Conversations on the New European Bauhaus, 2021 CONFERENCE. https://europa.eu/new-european-bauhaus/get-involved/events/conference-conversations-new-european-bauhaus-2021-04-22_en. Last accessed 27 June 2022
4. Smart Cities Marketplace: Integrated planning, policy and regulations. Action cluster roadmap 2021–2022. https://indd.adobe.com/view/publication/a87fdbf8-55f8-4f68-8bd6-96c6c6c0ab0b/fp8t/publication-web-resources/pdf/Roadmaps-Integrated_Planning,_Policy_and_Regulations.pdf. Last accessed 22 Nov 2022
5. European Commission: Clean energy for all Europeans package. https://energy.ec.europa.eu/topics/energy-strategy/clean-energy-all-europeans-package_en. Last accessed 27 June 2022
6. GECO project: Le comunità energetiche in Italia - Una guida per orientare i cittadini nel nuovo mercato dell'energia (2020). https://doi.org/10.12910/DOC2020-012
7. Schlindwein, L., Montalvo, C.: Accounting for heterogenous behaviours within transformative policy-making: the case of Energy Citizenship. In: EU-SPRI 2022 Conference, 1–3 June, Utrecht (2022)
8. European Commission: The European Green Deal, COM(2019) 640 final, Brussels (2019)
9. European Commission: 2030 climate & energy framework. https://ec.europa.eu/clima/eu-action/climate-strategies-targets/2030-climate-energy-framework_en. Last accessed 27 June 2022
10. European Commission: 2030 Climate Target Plan. https://ec.europa.eu/clima/eu-action/european-green-deal/2030-climate-target-plan_en. Last accessed 27 June 2022
11. European Union: Regulation (EU) 2021/1119 of the European Parliament and of the Council of 30 June 2021 establishing the framework for achieving climate neutrality and amending Regulations (EC) No 401/2009 and (EU) 2018/1999 ('European Climate Law'), Brussels (2021)

12. European Parliament: Fit for 55 package under the European Green Deal. https://www.europarl.europa.eu/legislative-train/package-fit-for-55. Last accessed 27 June 2022
13. European Commission. European Missions: 100 Climate-Neutral and Smart Cities by 2030. Info Kit for Cities 2021. https://ec.europa.eu/info/sites/default/files/research_and_innovation/funding/documents/ec_rtd_eu-mission-climate-neutral-cities-infokit.pdf. Last accessed 27 June 2022
14. European Commission: REPowerEU Plan. COM (2022) 230 final, Brussels (2022)
15. SmartEnCity project website. https://smartencity.eu/. Last accessed 22 Nov 2022
16. Covenant of Mayors for Climate & Energy: Reporting Guidelines March 2020
17. Covenant of Mayors for Climate & Energy homepage. https://www.covenantofmayors.eu/. Last accessed 27 June 2022
18. Maya-Drysdale, D., Krog Jensen, L., Vad Mathiesen, B.: Energy vision strategies for the EU green new deal: a case study of European cities. Energies **13**(9), 2194 (2020). https://doi.org/10.3390/en13092194
19. ISPRA: Stato di attuazione del Patto dei Sindaci in Italia. In: Rapporti 316/2020. ISBN 978-88-448-0988-1. ISPRA (2020)
20. Cassanmagnago, D.: Il Patto dei Sindaci. In: Il ruolo degli Enti Pubblici nel processo di transizione energetica: pianificazione energetica integrata, PAESC e comunità energetiche. SmartEnCity project training event of 27 April 2022. Lecce (2022)
21. Bologna Municipality: PAESC—Piano d'Azione per l'Energia Sostenibile ed il Clima. http://www.comune.bologna.it/paes/contenuti/143:47194/. Last accessed 27 June 2022
22. GECO Community. https://www.gecocommunity.it/partners/. Last accessed 27 June 2022
23. Ministero della Transizione Ecologica: Piano Nazionale Integrato per l'Energia e il Clima—PNIEC 2019; https://www.mite.gov.it/sites/default/files/archivio/pniec_finale_17012020.pdf. Last accessed 27 June 2022
24. Palermo, V., Bertoldi, P., Apostolou, M., Kona, A., Rivas, S.: Assessment of climate change mitigation policies in 315 cities in the Covenant of Mayors initiative. Sustain. Cities Soc. **60**, 102258
25. Boulanger, S.O.M., Massari, M., Longo, D., Turillazzi, B., Nucci, C.A.: Designing collaborative energy communities: a European overview. Energies **14**, 8226 (2021). https://doi.org/10.3390/en14248226
26. Roversi, R., Boeri, A., Pagliula, S., Turci, G.: Energy community in action. Energy citizenship contract as tool for climate neutrality. Smart Cities **5**(1), 294–317 (2022). https://doi.org/10.3390/smartcities5010018

Serena Pagliula, construction engineer, Ph.D. student in Architecture and Design Cultures at the Alma Mater Studiorum University of Bologna. She is in the TRACE research team of UNIBO, working on the H2020 GRETA—Green Energy Transition Action project. She was project manager of the H2020 SmartEnCity—Towards Smart Zero CO_2 Cities across Europe project for the Municipality of Lecce. She deals with topics related to the energy transition for climate-neutral and smart cities, with focus on energy plans integrating the energy poverty's issues.

Art of Temporary Living: Looking Inside Student Rooms

Arshia Eghbali

Abstract

Through four brief fragments, this chapter establishes a cause, an approach, and a context for an aesthetic inquiry into student rooms as places of transitory dwelling and materialisations of an art of temporary living. First, based on the notion of the "arts of doing" (*arts de faire*), the art of temporary living is defined as the ensemble of tactical, everyday practices that shape and are shaped by the students' landscape of home. Second, the infra-ordinary gaze is presented as an approach for investigating how the basic level of everyday life is lived through its mundane practices and materialities, Third, through a discussion of Georges Perec's novel *La Vie mode d'emploi*, and by situating it within the wider context of the visual culture of voyeuristic illustrations of urban domestic interiors, a certain aesthetic is defined for such illustrations along the lines of architectural precision coupled with an exhaustive attention to the mundane traces of lived space. Fourth, the student bedroom is discussed and studied through original illustrations created based on ads posted on online platforms for renting or sub-letting rooms to students in Bologna, Italy. Finally, two recurring material elements of the student room are explored in order to shed some more light on the art of temporary living as practiced by university students.

Keywords

Everyday life · Infra-ordinary · Voyeuristic gaze

1 Towards an Art of Temporary Living

From trying to find themselves a room in often-oversaturated housing markets and deciding what portions of their lives to fit in their suitcases, to decorating their rooms with small knick-knacks and negotiating space with roommates on a daily basis, many young people—and in particular, university students—are constantly dealing with the realities of temporary living. For both international and domestic students, moving to another location to pursue their studies entails living arrangements that are transitory and temporary in nature, such as living in dorms or shared apartments. This comes on top of the precariousness that young people are generally faced with in today's world, as well as multiple biographical transitions

A. Eghbali (✉)
Department of Architecture, University of Bologna, Bologna, Italy
e-mail: seyedarshia.eghbali2@unibo.it

Department of Arts and Cultural Studies, University of Copenhagen, Copenhagen, Denmark

A. Borsari et al. (eds.), *TEMPORARY: Citizenship, Architecture and City*, The City Project 4,
https://doi.org/10.1007/978-3-031-36667-3_10

—from youth to adulthood, from familial care to independence, and from studying to entering the job market, to name a few.[1] The ways in which students familiarise themselves with their new dwelling environments, take hold of them, and conduct their activities, together with the material setting of those living spaces weave the fabric of their everyday lives. It is the constellation of these practices and the different approaches, with varying degrees of success, to them that makes it possible to speak of a certain *art of temporary living*.

The *lived* quality of everyday life and its reliance on practices inspire a simple yet essential inquiry: how does one live? Or to be more precise: how does one live as an individual amidst all the various forces that constantly try to shape the everyday? Henri Lefebvre, one of the key thinkers of everyday life and the first to initiate a systematic analysis, puts the emphasis on the way we live: "what matters is not only what the social forces make of our everyday life but what we make of those forces through our ways of living" [5, p. 349, my translation]. An inquiry on how one lives, for Lefebvre, needs to encompass the "*art de vivre*" (art of living): "art of living presupposes that one considers one's life […] not like a means for 'another' end but as an end itself. It presupposes that […] everyday life becomes a work of art" [6, p. 213]. As such, human life, in the course of its everyday performance, is an object for aesthetic attention, not only in the narrower sense of the word that deals with matters of beauty, but also in the wider, original sense of the aesthetic as concerned with sensory perception. This view is also close to the notion of the art of living that is central to the field of Somaesthetics, as developed by Richard Shusterman, dealing with aesthetics in its wider sense and the soma as the sentient purposive body. The art of living, thus, is "an art necessarily performed with the soma" [7, p. 14]. The performative aspect of the art of living is at the very core of another seminal work on everyday life, Michel de Certeau's *L'invention du quotidien* [8]. Certeau argues for an essential inventiveness in everyday practices that enables the "ordinary" person (i.e., users and consumers) to go beyond what is expected of them and that which is dictated by design, policy, and production. Drawing on military jargon, Certeau distinguishes between "strategies" and "tactics" in the sense that he associates strategies with a totalising view when "a subject of will and power is isolatable from its environment", while a tactic is more spontaneous, as due to its "placelessness, a tactic depends on time, waiting observingly to seize passing possibilities of profit" [8, p. xlvi, my translation]. Tactics, therefore, are associated with doing, performing, and operating. In everyday life, we operate in a tactical manner; we invent our own ways of doing things within an imposed system. This is what Certeau calls the "*arts de faire*"[2] (arts of doing) [8]. Certeau's "arts of doing" are expressed first and foremost in one's landscape of home and familiarity—i.e., the primary spaces of the human body. The everyday practices of moving around the city, living in a neighbourhood, arranging one's furniture, or cooking are all tactical acts that reveal one's arts of doing.[3] And as Certeau's collaborator, Luce Giard writes, the ensemble of these arts of doing make up an art of living: "fragments of life whose secrets and poetic ruses wove the fabric of a soon-to-be lost time, ephemeral inventions of the obscure heroes of the ordinary,

[1] There has been a growing body of interesting literature on student homemaking in the past couple of decades since Liz Kenyon's groundbreaking study in 1999 on students' transitional experience of home [1]. A review of this expanding literature would be beyond the scope of this chapter. However, it is worthwhile to mention three studies that emphasise the framework of temporary living and everyday life: Rampazi's 2016 study on Italian students living temporarily in Northern Italy [2]; Pink and Postill's 2017 study on the mundane, everyday act of laundry within the transitory living arrangements of Indonesian students in Australia [3]; and Boccagni and Yapo's 2022 study on the temporary homemaking experiences of international students in Northern Italy as a form of dwelling in transition [4].

[2] L'invention du quotidien is in fact a two-volume publication. The first volume is sub-titled *arts de faire*, and the second volume *habiter, cuisiner*.

[3] Such concrete cases are explored in the second volume of *L'invention du quotidien*, which consists chiefly of two monographs based on interviews with people about their everyday practices: one by Luce Giard on cooking and the other one by Pierre Mayol on inhabiting a neighborhood [9].

arts de faire that make up without saying so an *art de vivre*" [9, p. xlv]. As such, in the case of students and their dwelling in transition, the art of temporary living refers to the many everyday practices, spaces, and objects that shape and are shaped by their landscape of home.

2 The Infra-ordinary Gaze

Now that we have established what an art of temporary living consists of, the main question regarding the study of students and their dwelling in transition is a somewhat methodological one: what to consider, where to look, and how to access this basic level of everyday lived experience? Everyday life is the realm of habits, routines, sameness, and repetition. The abundant and the mundane. But there lies the paradox of everyday life: it is both the most obvious and the most elusive. On the one hand, due to its repetitiveness and sheer closeness to our lived experience, whatever is considered "everyday" is by definition familiar. On the other hand, due to this very "everydayness", our most repeated actions and most inhabited spaces are the ones that are just taken for granted, and thus, escape systematic scrutiny. As Maurice Blanchot writes: "the everyday has this essential trait: it does not allow to be seized. It escapes. It belongs to insignificance" [10, p. 357, my translation]. Not in the sense that it is unimportant, but in the sense that it is an unmediated level of existence, where happenings and experiences are not yet assimilated into ready-made significations and abstractions. As such, everyday life is the non-negotiable reality that we live, ruled by practicalities, before it is loaded with significance.

This is where the notion of 'infra-ordinary' and George Perec's original approach come to our aid. Infra-ordinary (*l'infra-ordinaire*) was a neologism created by Georges Perec, Paul Virilio, and Jean Duvignaud,[4] which Perec elaborates in a short piece titled *Approches de quoi?* (Approaches to what?), published in 1973 in the fifth issue of their journal *Cause Commune*. Being the hands-on explorer he was, Perec's definition of the infra-ordinary is based on a simple dichotomy between that which is noticed and that which is generally neglected—the "event" versus "the rest", the "exotic" versus the "endotic". It is then the practical questions of what and how that direct his inquiry:

> What really happens, what we live, the rest, all the rest, where is it? What happens every day and recurs every day, the banal, the everyday, the evident, the common, the ordinary, the infra-ordinary, the background noise, the habitual, how to take account of it, how to investigate it, how to describe it? [12, p. 11, my translation]

Seeking to register *exhaustively* the infra-ordinary and to investigate how the basic level of everyday life is lived through its mundane practices and materialities, Perec's practical response is to turn the ethnographic gaze onto the things that are closest to us:

> Interrogate what seems so self-evident that we have forgotten its origin. [...]
> Describe your street. Describe another. Compare. Make an inventory list of your pockets, of your bag. Question yourself on the provenance, use, and what may become of each object you take out. Question your teaspoons. [12, pp. 12–13, my translation]

Perec creates a sort of estrangement mechanism to neutralise the cunning effects of a normalised everydayness. To avoid both hinders of ready-made signification and taken-for-grantedness. To investigate the familiar objects and practices of the everyday as they are, and not as they are reflected: "an effort to seize something that belongs to my experience, not on the level of its faraway reflections, but at the heart of its emergence" [13, p. 23, my translation] As such, the infra-ordinary gaze, in the Perecquian sense, is a practical approach—a *project*, which in Michael Sheringham's words is "partly ludic, partly tactical, always provisional and ultimately open-ended" [14, p. 199]. Just as for Certeau the everyday is invented through practices and arts of doing, for Perec the way to approach and

[4] In two separate interviews in 1991 with Andrea Borsari, both Virilio and Duvignaud confirm the creation of the neologism and the development of the concept as a project between the three of them [11].

investigate those arts of doing, or an art of living for that matter, is an inventive practice itself. The infra-ordinary gaze is the art of investigating the art of living.

3 A Certain Kind of Voyeurism

Reflecting on the infra-ordinary, Virilio speaks of a certain "gaze that, in fact, Perec and I exercised with more efficiency; we were more discerning *voyeurs* that others" [11, p. 264, my translation]. To frame the infra-ordinary perspective as a voyeuristic gaze is indeed illuminating. The infra-ordinary ran through Perec's works (both fiction and non-fiction) like a thread, but it is arguably epitomised in his most celebrated novel *La Vie mode d'emploi* (Life A User's Manual) [15]. In *La Vie mode d'emploi*, Perec imagines a Parisian apartment building whose façade is lifted, and all the pieces are rendered visible in one moment of time. Each of the 99 chapters is the exhaustive description of one of these interiors. Employing the infra-ordinary gaze, Perec lists and describes all the negligible knick-knacks, small objects, furniture items, and the persons that are under this voyeuristic gaze. But as the provenance, use, and fate of each object and space gets explored, narratives multiply and transgress the limits of the room's time and space, creating a massive, intricate puzzle of interior scenes and narratives. The pictorial logic of the novel is accentuated by Perec in a ludic way as we discover, within the book, that this thorough and exhaustive dissection of the building is in fact the project conceived and pursued by a painter character in the novel, an inhabitant of the building. This graphic dimension of Perec's mammoth "architext" [16] relates in more than one way to the wider context of an existing visual culture of somewhat voyeuristic depictions of urban domestic interiors. There are several (visual) sources of inspiration for *La Vie mode d'emploi*; as Perec himself has clearly stated in a short passage in his 1974 book *Espèces d'espaces* (Species of Spaces) [17], in which he lays out his plan for the novel. In the same text, Perec cites two specific sources: Alain-René Lesage's 1707 novel *Le Diable Boiteux* [18] and Saul Steinberg's illustration known as "No Vacancy" from his 1949 collection *The Art of Living* [19], which is a fine-line drawing of a New York rooming-house whose façade is lifted and the interiors are revealed, where we can see the inhabitants in their everyday lives. Among Perec's other visual sources of inspiration,[5] one can find antique doll houses (*maisons de poupées*), a drawing by Jean-Jacques Sempé, A photomontage by Robert Doisneau titled "*La Maison des locataires",* and an article on Nineteenth-century illustrations depicting cross-sections of Parisian buildings and the lives of their inhabitants, accompanied by reproductions of some of the illustrations such as those published in the 1847 [21] and the 1883 [22] issues of *Le Magasin pittoresque* (see Fig. 1).

Nineteenth-century Paris saw a boom of writings and imagery, in illustrated press, fascinated by depicting an exhaustive overview of the multiplicity of everyday life among city dwellers. In *The Arcades Project*, Walter Benjamin termed it "panoramic literature" [23]. The panoramic texts and images strived to give objective details about the lived complexity of everyday life in everyday spaces and, narrative-wise, were composed of micronarratives that were not necessarily connected [24]. *Le Diable boiteux* was a highly influential precedent for both its literary and visual legacies. Based on an earlier seventeenth-century Spanish novel, it recounts the story of a devil that in order to thank the man who freed him from a bottle (a young student, in fact!), flies him above Madrid and lifts the roofs of the buildings so that he can see the interiors and the people inside without being seen. Moreover, the 1726 edition of the novel features illustrations by François-Philippe Dubercelle, which depict buildings whose roofs and walls have disappeared and the interior spaces and the people inside are revealed. The novel and the

[5] Based on materials from Perec's manuscripts and scrapbooks, Danielle Constantin [20] studies some of the lesser-known sources that inspired *La Vie mode d'emploi.*

Fig. 1 "*Paris qui travaille*" (Paris working), by Tissandier and Gilbert, 1883. *Source* Le Magasin pittoresque 51. Edited by E. Charton. Paris, 1883

illustrations present a few distinctive elements: the *coupes habitées,* or the slices of life in cross-sections of apartment buildings; the curiosity about how everyday life is lived in private interiors; a rather technical, architectural way of illustrating the interior spaces through section cuts (they almost look like axonometric section drawings produced by architects today); and a certain kind of voyeurism that is even more accentuated via the figure of the devil. The novel and the illustrations inspired a whole set of "diable-themed" works in the 1800s that made up an important part of the panoramic literature of the time [25]. One example that will bring us all the way back to Perec's *La Vie mode d'emploi* is Bertall's 1845 illustration titled "*Coupe d'une maison parisienne le 1er janvier 1845—Cinq étages du monde parisien*" (The cross-section of a Parisian apartment building on 1 January 1845—Five levels of the Parisian world), which appeared in the publication *Le Diable à Paris: Paris et les Parisiens* (The Devil in Paris: Paris and the Parisians) [26]. The illustration that depicts domestic life as seen through the cross-section of a five-story apartment building with technical precision has long served as the artwork on the cover of the widely available paperback edition of *La Vie mode d'emploi* [15].

The rich visual culture of the voyeuristic gaze into the interiors of the modern city provides a historical context and an aesthetic grounding for visual explorations of everyday life, or as it is in the title of Steinberg's collection of drawings—which resonates perfectly with the general theme of this text—the “art of living”. The defining aesthetic elements of these kinds of illustrations are a certain architectural precision coupled with an exhaustive attention to the mundane traces of lived space. This precedence is particularly interesting in today’s world of hypervisuality and excessive imagery, where we are experiencing an enormous surge in the number of glimpses that we get into other people’s private spaces, through social media, livestreaming, and video calls (all of which of course received yet another boost during the COVID-19 pandemic). And since a large number of the users of these outlets and networks are young people, they are broadcasting from their bedrooms,[6] putting the intimate space of the bedroom out there.

4 Looking Inside Student Rooms

One of the ample sources for images that depict contemporary domestic interiors—and bedrooms in particular—are the ads posted on online platforms for renting or sub-letting rooms to students. These platforms are the go-to means for students who search for accommodation or those who look for roommates to join them. A student room is but an augmented bedroom. A bedroom that is supposed to be the gravity centre of the student’s landscape of home. The transitory nature of student life means that these rooms are illustrative examples of temporary living spaces, and thus, embodiments of the everyday aesthetics of temporary living. Looking inside these spaces can provide us with valuable insights into the art of temporary living as it is practiced by students. But first, it is worthwhile to have a closer look at the bedroom as a kind of space, as an object: what is a bedroom?

The bedroom is the basic spatial-architectonic unit of habitation, commonly symbolising privacy, intimacy, and withdrawal.[7] Like the kitchen and the bathroom, the bedroom is also a part of the home that is closely tied to the human body and primary bodily functions. Its main purpose is to provide a space for sleep, rest, and repose. Materially, a bedroom, at its most fundamental level, is composed of the basic elements that constitute a room—i.e., walls, floor, ceiling, window, and door—plus an essential piece of furniture, which has given its name to it: *the bed.* Other items may be present, too. But the sole defining feature of the bedroom remains the bed, which itself is a unique kind of space, characterised by the dimensions of the human body. The space in which, contrastingly to nearly all other spaces and situations, the human body is in a horizontal position: “the individual space par excellence, the elementary space of the body”, as Georges Perec writes [17, p. 26]. As a place, the bedroom has different meanings for its many different occupants. For young children, old adults, and the sick, it is a place of care. For teenagers, it is a place to assert their evolving identities, the closest they have to a personal space in the familial home. And for precarious young people, students, exiles, and immigrants, it is a “room of one’s own”—an equivalent of home within larger, more precarious, shared living arrangements. As such, the materialities of the bedroom and the everyday practices that take place within its walls provide a unique empirical field. The inclusions and exclusions, preferences, the extent to which one is granted control over the space, arrangements of the furniture and personal belongings, manners of organising the space, order or disorder, and routines compose personal and socio-cultural narratives that are embedded in the bedroom.

The centrality of nocturnal rest and its materialisation, the bed, make the bedroom an indispensable locale for any type of sojourn, no matter how short or long. Lodgers, students, tourists and

[6] See Blanchfield and Lotfi-Jam’s 2017 paper [27] on how objects in the bedroom are used to curate a certain image of the space in order for it to be displayed online; and Ruberg and Lark’s 2021 paper [28] on bedrooms as seen in the background on the popular livestreaming platform Twitch.

[7] It is worth noting that, as historians tell us, our current notions of privacy are fairly new [29, 30]. It was not uncommon for bedrooms and even beds to be shared by multiple persons for sleep, or for guests to be entertained in bedrooms (as many students do now, since they do not have another living space at their disposal).

business travellers all need a place to spend the night. It comes as no surprise that the bedroom is the model for various forms of transitory accommodation: hotel rooms, dorm rooms, bedsits, etc. For the temporary dweller, the bedroom is a practical necessity, but also a reference point from which to navigate everyday life and relate to other spatial scales. Mapping the bedroom of the temporary dweller provides a unique vantage point to study the complex and multi-scalar dynamics of belonging and homemaking, not only limited to the interior of the bedroom, but also in relation to the city and beyond.

From December 2020 to February 2021, as part of my larger research project on students' temporary homemaking, I surveyed ads posted on Facebook for student rooms in Bologna, Italy, and eventually selected 10 of the rooms to recreate through 3D modelling, based on the uploaded photos and the given descriptions in the ads. Then, based on the 3D models, I created two sets of drawings from the rooms: axonometric drawings depicting the room in its entirety (see Fig. 2) and perspective illustrations of the space of the rooms (see Fig. 3). The outcome is a series of snapshots of students' everyday domestic interiors. The infra-ordinary gaze, which is amplified even more through the meticulous process of reconstructing these spaces, allows to recognise and analyse the material elements of the room that represent the art of temporary living. These elements also point to other temporal and spatial scales that go beyond the actual time and space of the room (e.g., a postcard on the wall is not only a decorative element, but it also connects the space of the room with the occupant's memories, faraway locations, and past experiences). As such, this also underlines the synecdochical quality of the room as the centre of the temporary dweller's landscape of home. In

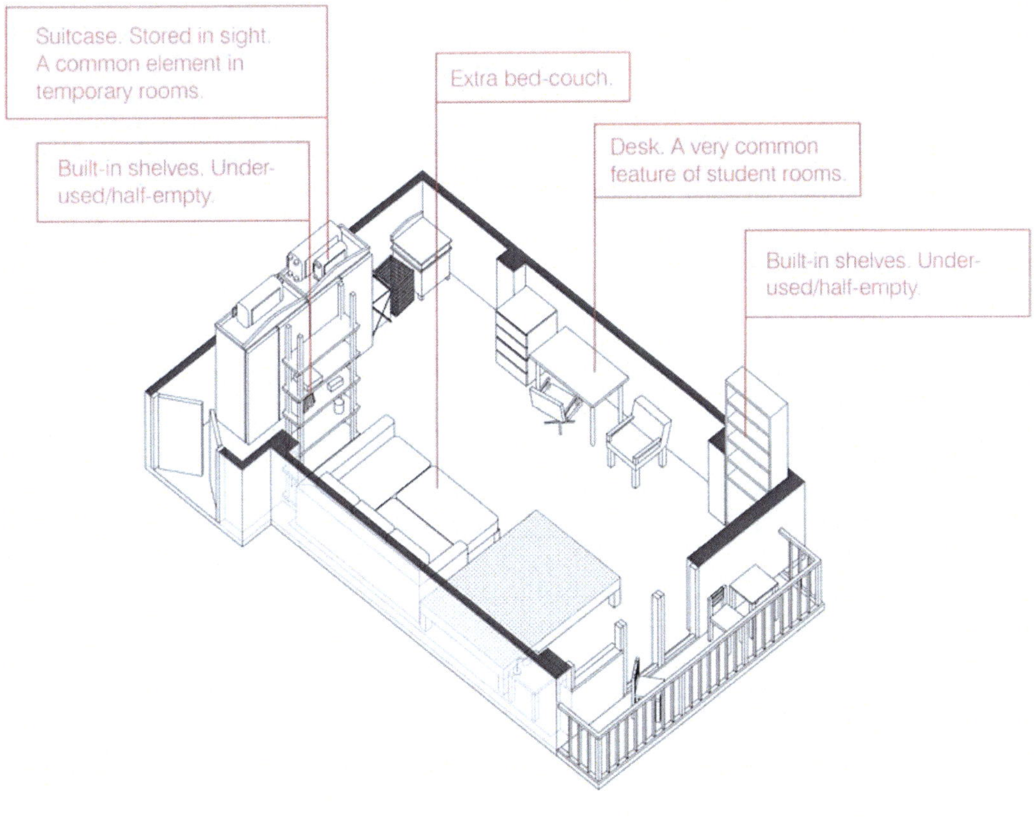

Fig. 2 Axonometric drawing of a student room in Bologna, Italy, 2021. *Source* Author

Fig. 3 Perspective illustrations of two student rooms in Bologna, Italy, 2021. *Source* Author

what follows, as a conclusion to this chapter, we look at two such elements that were often present in the student rooms analysed.

4.1 The Suitcase

The suitcase is perhaps the most iconic object associated with living in a transitory living arrangement. It contains and makes possible the mobility of one's primary material belongings. The suitcase, as such, stars twice in the home-making experience of students, as temporary dwellers, in every new home: once upon moving in and once upon moving out. But it also has a continuous presence in their homely worlds. In most rooms it is visibly present, either in a corner or on top of a closet, or in some other place in sight. In most of these interiors, the suitcase, which is not part of the furniture of a conventional home, has acquired the status of an item of furniture. Its relationship with the room and its inhabitant—like any object in an interior—is threefold. The first layer is simply dictated by the materiality of the room and the object itself. The physical layout of the room and the arrangement of its furniture, coupled with the shape and dimensions of the suitcase, limit its potential position, so it ends up where it is placed. The second layer deals with the function that the object adopts due to its new position, which can be different from its initial purpose. For example, it can be used as extra storage space, or even almost like a cupboard. This layer highlights the tactical aspect of temporary homemaking. The third layer is a symbolic one. But this symbolic value is not the result of some a priori meaning invested in the object (unless one makes the active decision to put a suitcase in an interior, merely as decoration or a work of art). It is, instead, the integration of the suitcase in the room that creates a setting which gives a *temporary* form to the inhabitant's everyday life. That is to say, keeping a big suitcase in the middle of a rather small room, or using it as a cupboard is not the most sustainable arrangement. This echoes Daniel Miller's concept of the reciprocity of the relation between human beings and material objects, in that they both mutually define each other and that our material surroundings set the setting for our behaviours

[31]. The temporariness of the student's living arrangement brings in and maintains the suitcase where it is, while the suitcase, by being there, suggests and confirms that this arrangement is temporary.

4.2 Pictures, Postcards, and Posters

One of the challenges of dwelling in transition for students is the limited resources that they have or are ready to invest in their living spaces. On top of that comes the lightness that is associated with temporary living; if they are supposed to fit their life in a suitcase or two, then there is little room left for carrying large items of personal value. A very common sight in student rooms are pictures, postcards, and posters, the tactical response of the temporary dweller to various problems, from the lack of personal identity and homely feeling, to simply the blankness of an otherwise all-white room. Even if not for reasons of profound personalisation and identity-building, they remain the quintessential objects of temporary homemaking. They are easily portable, versatile, and collectible. More than anything, they are objects of a mobile lifestyle; they are usually collected through traveling, moving, and maintaining relationships with distant people and places. As such, these small, lightweight objects can bring a myriad of spatial and temporal scales into the student's room.

References

1. Kenyon, L.: A home from home: students' transitional experience of home. In: Chapman, T., Hockey, J. (eds.) Ideal Homes? Social Change and Domestic Life, pp. 84–95. Routledge, London and New York (1999)
2. Rampazi, M.: Temporary homes: a case study of young people's dwelling strategies in northern Italy. Space Cult. **19**(4), 361–372 (2016)
3. Pink, S., Postill, J.: Student migration and domestic improvisation: transient migration through the experience of everyday laundry. Transitions J. Transient Migr. **1**(1), 13–28 (2017)
4. Boccagni, P., Yapo, S.: International students and homemaking in transition: locating home on the threshold between ascription and achievement. Home Cult., 1–19 (2022)
5. Lefebvre, H.: La Vie quotidienne dans le monde moderne, p. 376. Gallimard, Paris (1968)
6. Lefebvre, H.: Critique de la vie quotidienne I: Introduction, p. 272. L'Arche, Paris (1958)
7. Shusterman, R.: Bodies in the streets: the soma, the city, and the art of living. In: Shusterman, R. (ed.) Bodies in the Streets: The Somaesthetics of City Life, pp. 13–37. Brill, Leiden and Boston (2019)
8. Certeau, Md.: L'invention du quotidien: 1. arts de faire, p. 350. Gallimard, Paris (1990)
9. Certeau, Md., Giard, L., Mayol, P.: The Practice of Everyday Life. Volume 2: Living and Cooking. University of Minnesota Press, Minneapolis (1998)
10. Blanchot, M.: L'entretien infini. Gallimard, Paris (1969)
11. Borsari, A.: Le cause communi. Una rassegna e tre conversazioni intorno a Georges Perec. Nuova Corrente **108**, 249–96 (1991)
12. Perec, G.: L'Infra-ordinaire. Seuil, Paris (1989)
13. Perec, G.: Penser/Classer. Hachette, Paris (1985)
14. Sheringham, M.: Attending to the everyday: Blanchot, Lefebvre, Certeau, Perec. French Stud. **54**(2), 187–199 (2000)
15. Perec, G.: La Vie mode d'emploi. Fayard, Paris (1978/2010)
16. Mitchell, P.: Constructing the Architext: Georges Perec's life a user's manual. Mosaic Interdiscip. Crit. J. **37**(1), 1–16 (2004)
17. Perec, G.: Espèces d'espaces. Éditions Galilée, Paris (1974)
18. Lesage, A.R.: Le Diable Boiteux. Chez la veuve Barbin, au Palais, sur le perron de la Sainte Chapelle, Paris (1707)
19. Steinberg, S.: The Art of Living. Harper and Brothers, New York (1949)
20. Constantin, D.: Les maisons de poupées de Perec: un catalogue. In: Burgelin, C., Heck, M., Reggiani, C. (eds.) Cahier de L'Herne N°116: George Perec, pp. 260–266. L'Herne, Paris (2016)
21. Charton, E. (ed.): Le Magasin pittoresque 15. Paris (1847)
22. Charton, E. (ed.): Le Magasin pittoresque 51. Paris (1883)
23. Benjamin, W.: The Arcades Project. The Belknap Press of Harvard University Press, Cambridge and London (1999)
24. Cohen, M.: Panoramic literature and the invention of everyday genres. In: Charney, L., Schwartz, V. (eds.) Cinema and the Invention of Modern Life, pp. 227–252. University of California Press, Berkeley (1995)
25. Willems, P.: Virtual reality in the age of panoramas: mapping out buildings, a village, capitals, and hell. Dix-Neuf **20**(2), 187–212 (2016)
26. Balzac, Hd., et al.: Le diable à Paris: Paris et les parisiens. J. Hetzel, Paris (1846)
27. Blanchfield, C., Lotfi-Jam, F.: The bedroom of things. Log **41**, 129–134 (2017)

28. Ruberg, B., Lark, D.: Livestreaming from the bedroom: performing intimacy through domestic space on Twitch. Convergence Int. J. Res. New Media Technol. **27**(3), 679–695 (2021)
29. Rybczynski, W.: Home: A Short History of an Idea. Penguin Books, London (1986)
30. Bryson, B.: At Home: A Short History of Private Life. Doubleday, London (2010)
31. Miller, D.: Stuff. Polity, Cambridge (2009)

Arshia Eghbali is a Joint Ph.D. Candidate at the Department of Architecture, University of Bologna, and the Department of Arts and Cultural Studies, University of Copenhagen, conducting his research on temporary home cultures, everyday life, and Georges Perec. Arshia holds a Joint Master's Degree in Urban Studies (4CITIES) from Brussels, Vienna, Copenhagen, and Madrid, and a Bachelor's in Architecture from University of Tehran. His work consists of research, process-oriented art, design, writing, curating, and independent publishing; some of which have been exhibited in Tehran, Stockholm, Copenhagen, and Rotterdam.

Permanent versus Temporary: A Struggle within City Transformations

Annalisa Trentin

Abstract

Permanence and temporariness are two counterposed terms that characterise architecture and urban transformations. This essay intends to address some of the many interpretations of this binomial: from the role of urban permanence and the temporaneity of use, as addressed by Aldo Rossi in *The Architecture of the City*, to the concept of the temporary duration of the life of buildings and the concept of conservation, presented by Rem Koolhaas with his research *Cronocaos*. Next, the fundamental role of unplanned areas is analysed as a place for experimentation of the temporary, as a means to test future scenarios, taking as an example the survey conducted by the Urban Catalyst group, and then concluding with the experience of the IPA—Institute for Public Architecture, in the desire to reduce the transitory nature of living, avoiding gentrification and offering a permanent living condition to the most disadvantaged social groups. The examples presented are intended to illustrate how architecture and the city often live in the balance between permanence and transience, that there are no absolute instances and that only a well-structured architecture, on a formal basis that refers to a collective memory, can embrace the continuous transience of function and always remain vital within the urban organism.

Keywords

Permanence · Temporariness · Architecture

The temporary/permanent binomial marks the very essence of architecture and the city, whether buildings or portions of the city, a conflict that represents a crucial point in urban transformation processes. The city is by its nature structured by a continuous antithetical relation between permanent and temporary, where the myth of permanence clashes with the fact that the city is a dynamic entity where the crystallisation of the status quo is to be considered pure utopia.

1 Permanence of Form/Temporariness of Function

The literature related to urban studies, the nature of cities and their transformations in history were brought to synthesis by Aldo Rossi in the essay

A. Trentin (✉)
Department of Architecture, University of Bologna, Bologna, Italy
e-mail: annalisa.trentin@unibo.it

A. Borsari et al. (eds.), *TEMPORARY: Citizenship, Architecture and City*, The City Project 4,
https://doi.org/10.1007/978-3-031-36667-3_11

The Architecture of the City,[1] where from the first chapter, devoted to the structure of urban facts, he introduces the concept of function, mentioning the rich and articulate conception given by French geographer Georges Chabot.[2] While Chabot classifies the main functions of the city into military, commercial, therapeutic, intellectual, religious, and administrative, he also states that these functions are transient and elementary rather than permanent facts in the city: for Chabot, function is a moment of urban life, thus assuming a value of temporariness. Rossi, in the introduction to his treatise, criticizes naïve functionalism; the critique of functionalism should for Rossi be seen as a new theory of architectural composition, reiterating the concept that function is not permanent. Function is understood in an algebraic sense, which means, as Rossi himself writes, that its values are knowable one in function of the other, and that more complex links are established between function and form than the linear ones of cause and effect, which are then contradicted by reality. The permanence of function, expressed through regulatory instruments and the zoning of urban plans, becomes a principle of simplification of reality and humiliation of imagination and freedom.[3] For Rossi, the layering of the contemporary city highlights the relation between the permanence of traces, the foundational elements of urban structure, the presence of monuments and at the same time with its constant changes, with fluctuations dictated by the economic and social context and the changing needs, where the obsession with the permanence of the plan has given way to a quest toward flexible planning, capable of responding to temporary needs. The demand for new uses has given rise to the development of strategies for structuring more flexible plans that have given rise to temporary use initiatives to meet ever-changing needs.

Non-permanent solutions have always been perceived as secondary because dictated by uncertainty, by the awareness of their transience, in contrast to a Western culture based on the concept of permanence. Aldo Rossi's essay *The Architecture of the City* showed how, through a different perception between past and present, it is possible to recompose the picture of an urban culture. Referring to the concept of permanence proposed by Marcel Poëte, Rossi brought attention to urban facts and on how the present helps us to understand the strong elements of the past, the generating elements of urban form.[4] Rossi, while recognising the value of permanence of monuments, addresses his view to the discontinuity of use, to the impermanence of function, and to the vitality of these primary elements.[5]

The permanence of form and the transience of use, the adaptation of use, is clearly expressed by Marcel Poëte when he states that one cannot fully grasp the meaning of urban life if one does not distinguish the "urban being", constituting—in itself—the social aggregate that essentially forms the city, from the "urban form" with which the city presents itself materially to our eyes: "Now it is to the urban being that biological laws need to be applied. Form is nothing more than the inert matter, shaped or arranged by the being and which, therefore, is not to be confused with. Adapted to its needs when it was created, this form only corresponds to it imperfectly when subsequent generations use it, still preserved, as there is a permanent background. The successive generations that make up the urban being succeed one another and it is the form—which remains—that expresses the urban soul".[6]

In 1968, in his first Blue Notebook, No. 1 dated 19 June, Aldo Rossi elaborates the concept of form: "If there is the possibility of a series of operations on architecture, these can take place on the basic form" and "the use of the building is the task of architecture",[7] to reiterate how the formal, the typological structure, embodies the theme of permanence, but opens the space to a transience of its use. Rossi points out how architecture is characterised by a

[1] Rossi [1].

[2] Chabot [2].

[3] Rossi, op. cit. [1], p. 7.

[4] Manzione [3], p. 174.

[5] Rossi, op. cit. [1], p. 51.

[6] Poëte [4].

[7] Rossi [5].

continuous change in the use of architectural type, as a logical form is connected to use and not to a predetermined function. The type, thus, represents an element of permanence of a formal structure, classifiable, categorisable, independent from a function and related to a use that can change over time and then is able to represent a character of temporariness.

With *The Architecture of the City*, Rossi attempts to define the points that characterise architectural type, stating that type is constituted by the needs and aspirations for beauty; is unique and yet very diverse in different societies; is related to the form and the way of life; it is something permanent and complex; it is a logical statement that precedes form and constitutes it.

The topic of permanence is investigated by Rossi through a careful analysis of monuments, as examples of collective memory, as primary elements capable of accelerating the process of urbanisation of a city and, referring to a wider territory, elements that characterise the processes of spatial transformation of the same territory. The type governs monuments and primary elements, determining their permanence and continuity with history, but at the same time allowing a continuous adaptation to reality and to the needs with which it is dealing. Persistences, according to Rossi, are detectable precisely through the monuments, the physical signs of the past, but also through the persistence of the layouts and the plan, and again he recalls Marcel Poëte who noted how the urban facts are sometimes endowed with a continuous vitality, sometimes they fade away and only the form, the physical sign, of the *locus* remains.[8]

This vision brings attention to the urban structure that is characterised by urban facts, monuments and primary elements that represent the topic of permanence, memory, and historical continuity and can accommodate the transformations and transitory needs dictated by reality. The areas, defined by Rossi as residential areas, constitute a kind of plain ground, and assume a secondary value, even if relevant, regarding the complex combination of the city with the coexistence of urban facts and residential areas.

2 Monuments, Permanences and a Building's Lifespan

Today we are witnessing a reversal of the point of observation. The monuments and the established city take a subordinate place to the surroundings, to everything that is not clearly definable as a drawn and recognisable part of a city, but that is for all intents and purposes part of the urban organism. Now, the space between established urban structures and the so-called minor architecture, meaning, for example, social housing and industrial buildings, becomes the subject of urban transformations, driven by new needs, and it is precisely through the network between formal and informal architecture and use of space that new strategies for the contemporary city come to be defined.

In the 2010 Architecture Biennale, Rem Koolhaas with his exhibition *Cronocaos* shifted the focus to secondary public heritage, particularly social architecture, at a time when legislation on permanence and preservation had defined rules that discriminated against some categories of buildings whose role in the urban structure was certainly not secondary. What are the rules that define the lifespan of an architecture? Everything is focused on preservation, but for Koolhaas it is at the same time necessary to make an argument about architecture that no longer plays any role in the urban organism and can in a sense be erased: "There is an over accelerating rate at which architecture has been declared as an item of permanence". From this, Koolhaas extrapolates that the next step, prospective preservation, would presumably predefine a building's status as monument, even before it is physically manifested. In response, Koolhaas presented a theory of its opposite, not what to keep but what to erase and abandon.[9] "In a global groundswell of revulsion, one particular genre

[8] Rossi, op. cit. [1], p. 52.

[9] Bishop [6], p. 15.

has escaped the embrace of preservation. Open season has been declared on postwar social architecture. At its zenith, a strong public sector created the conditions in which architecture as a social project could flourish. At its nadir, a public sector, debilitated by the market, destroys it. There is now a global consensus that postwar architecture—and the optimism it embodied about architecture's ability to organise the social world—was an aesthetic and ideological debacle. Our resignation is expressed in the flamboyant architecture of the market economy, which has its own built-in commercial expiration date".[10]

If monuments, or what is called monument by regulations, represent the myth of permanence, social architecture, such as postwar architecture, is not considered, thus it represents architecture with a finite time and is therefore temporary. Koolhaas notes how preservation requires the development of a theory of the opposite: not what to preserve, but what to surrender, what to erase and abandon.[11] In the accompanying report to the exhibition, Koolhaas addresses the critical issues, ambiguities and contradictions related to the issue of building preservation by stating that the criteria for selecting buildings for preservation are by definition vague and elastic, because they must embrace all the conditions the world contains; he also notes that time cannot be stopped, but in the "arsenal of preservation" there is no consideration of how to manage its effects, how the preserved can remain alive and at the same time evolve, transform, and change their function. Koolhaas points out that in conservation there is little awareness about how different cultures have interpreted permanence, or variations in materials, climate and environment, which in themselves require radically different modes of conservation, and that with its unspoken ideology, conservation prefers some authenticities above others. Other buildings—typically those that are politically difficult—are suppressed, even though they are crucial to the understanding of history. The continuous emphasis on the preservation of the exceptional —that which is worthy of preservation—creates its own distortion. The exceptional becomes the norm. There are no ideas to preserve the mediocre, the generic, to give permanence to common buildings,[12] but a city is not made only by monuments. The generic, the incomplete, the transient, and the vague become part of the urban organism. Their use, even temporary, can be the impetus for a process of urban development related to the redevelopment of urban areas.

3 Informal Processes of Temporariness

Cities have at their disposal an enormous number of vacant buildings and land. Although these are potentially attractive in the long term as sites for urban development, their currently indeterminate and temporary nature is precisely what makes them fertile ground for a wide range of creative uses. Vacant areas, parking lots, vacant buildings, and unplanned spaces are now quickly becoming particularly vibrant places in many cities: clubs, bars, start-ups, arts scenes, migrant economies and informal markets, recreation and nightlife take up residence in these places that urban planning and the real estate market initially fail to develop. This is often where innovative cultural productions and vital public spheres are found. Temporary use of these spaces can be the mechanism for understanding the needs and their potential for development and become the point of observation on how society evolves, not to satiate a market will, but as an act of care.[13]

In this shift of perspective, the focus moves from form to use; it is use that determines quality; the type of use, even temporary, is what makes vital a monument, an architecture or even a forgotten portion of the city. The extent to which temporary users can be consciously placed in spaces for which there is still a lack of medium-term development prospects—almost as agents of revitalisation—was investigated by a

[10] Koolhaas [7], p.122.

[11] *Ivi,* p.123.

[12] https://www.oma.com/projects/venice-biennale-2010-cronocaos.

[13] Borasi [8], p. 7.

research project commissioned by the Berlin Senate and a symposium held with representatives of the city administration, real estate agents and temporary users, followed by an exhibition on Space Pioneers in 2007. Through the research, exhibition and publication entitled *Urban Pioneers*,[14] a survey was conducted where one hundred different types of temporary use of vacant sites were mapped, models were created to demonstrate the potential of temporary use for urban development, and guidelines for temporary users were established.

This research has given rise to ongoing insights into the new uses of these areas that elude precise classification. In particular, the Urban Catalyst research group coordinated by Philipp Oswalt, Klaus Overmeyer and Philipp Misselwitz has devoted several years to these unplanned phenomena in five European countries, analysing their hidden logic. Based on its own involvement in several European-scale projects, Urban Catalyst has described the ways in which informal processes can find their way into urban planning and what urban planners can learn from intermediate users.[15] Their study has focused on understanding the temporary uses, of unused spaces within cities. The topic is treated in a very pragmatic way; there are no references to architectural composition, instead economics, laws and politics are discussed. Philipp Oswalt had already carried out research on the development of the city of Berlin, published in the volume *Berlin Stadt ohne Form: Strategien einer anderen Architektur*[16] where had analysed urban development and had openly criticised the work of Hans Stimmann, senate building director for urban development in Berlin, and the planning developed under his leadership in the 1990s. His research had focused not on monuments or the form of the city, but on the free spaces between buildings, investigating possible transformations and uses, including the different natures of temporary use. In 1998, speaking of Berlin as a 20th-century city, he wrote that "The central themes of the city: emptiness, fragmentation, heterogeneity, multiplicity, temporariness, formlessness, and subversion reveal a high degree of innovative potential. It remains to be hoped that, despite all trends toward restoration and economic exploitation, the city does not fully return to normal, but retains its particular identity and uses this to develop potential which points the way for the future".[17]

We are facing a further reversal, the priority is no longer the formal structure of architecture, the element of permanence that can accommodate different uses, but the formless space that can take on new meaning through use. The form of architecture takes a back seat to let users transform and informally plan the space. The research developed by the Urban Catalyst group, on *The Power of Temporary Use*, was able to highlight the various ways in which urban planning could incorporate informal and temporary processes, analysing them as an important lesson for understanding the changes in contemporary society. Temporary use of space has been classified through a typological matrix that can give structure and order to the different possibilities of use, a classification which is useful for replicating different strategies and to give them a name.

In the chapter Patterns of the Unplanned,[18] the authors have identified nine different types of temporary use of space and have assigned to each of them a name that can describe the action: STAND IN, which takes advantage of the lapse of time between the previous use and the next use; FREE-FLOW, where the function continues but is moved to new areas as opportunities arise; IMPULSE, meaning new use of a space that results in its new characterisation; CONSOLIDATION, where temporary use becomes time-limited and is regularised with permits; CO-EXISTENCE, where two different uses coexist because one of them remains of niche; PARASITE, where temporary use demonstrates the possibility of indefinite use; PIONEER, where a place used is transiently used for another activity, but that activity is successful and is thus carried

[14] Lauinger et al. [9].

[15] Oswalt et al. [10].

[16] Oswalt [11].

[17] Oswalt [12].

[18] Oswalt et al. [13], pp. 35–51.

on over time; SUBVERSION, where temporary use strategically establishes itself in the spaces of an ongoing use and subverts it; and DISPLACEMENT, where permanent use is temporarily moved to another place and then relocated. This change generates an impulse for renewal of activities.

This mode of temporary use of space consistently guides new planning, becomes a tool to prevent the constant process of gentrification in contemporary cities with transformation processes driven 'from below' and not imposed for purely speculative matters dictated from above by market laws.

4 Against Temporariness

Another example of the clash between temporary and permanent practices in the city is a countercultural one, presented as part of the international seminar *TEMPORARY Citizenship, Architecture and City*,[19] where activist Monxo Lopez, architect Jonathan Kirschenfeld, member and founder of IPA—Institute for Public Architecture, and architecture historian Susanne Schindler brought attention to the case study of New York with a presentation titled Mott Haven: Birthplace or Graveyard of Planning Ideas?

They focus their intervention on the need to avoid temporary accommodation and instead find permanent solutions for the inhabitants of the Mott Haven district. Today, as Susanne Schindler wrote in an article titled *Model Conflicts*, the main concern of residents of low-income neighbourhoods is displacement not through disinvestment, but through gentrification. This struggle can be seen through some of the areas targeted by the Model Cities survey half a century ago, including Mott Haven in the Bronx, East Harlem in Manhattan, and East New York in Brooklyn, but the city mayor's response is towards more housing construction rather than a comprehensive approach to address inequality.[20] There is a general search for stability, for the elimination of inequalities, for permanence, and there is now a focus on the removal of conditions of temporariness and uncertainty, especially in areas that after years of disruption, degradation and crime, have begun to recover, is very high.

In an interview with the New York Times,[21] Monxo Lopez said that gentrification had brought him to the Bronx from Brooklyn after his neighbourhood of Fort Greene had changed dramatically. Local stores and restaurants had closed, replaced by pet stores and boutique pizzerias, while corner delis had become more upscale, and the face of the neighbourhood had changed. Monxo Lopez, who is among the founders of South Bronx Unite, an activist group that is challenging the speed of development, also noted that everything was becoming more fragmented, and the neighbourhood's atmosphere was being lost. In 2006, real estate guru Barbara Corcoran decreed that the South Bronx was among the top five most desirable real estate markets in the nation, saying that public money was flowing in, that the arrival of artists and musicians could improve the area, and that the opening of expensive bars and clubs was a sign of an area on the upswing. Lopez said that fear then turned to outrage when developer Keith Rubenstein announced the construction of luxury towers in Port Morris, an area he tried to rename "The Piano District"; there was the idea of rebranding the area, prompted by the arrival of boutiques, cafes, and new temporary businesses, with the neighbourhood branding as a means for speculative activity. Lopez saw these operations, imposed top-down, as an effort to erase the history of the neighbourhood and the cultural and social achievements of a community, in favour of speculation. The residents of the neighbourhood were not involved or considered, now all these new businesses are at the centre of attention and the ones that have served the community are ignored and the history of the community, through good times and bad, has been erased. Lopez hopes development can be slowed and he has embraced the idea of a Community Land Trust to obtain city-owned buildings or lots that

[19] Ascari et al. [14].

[20] [15].

[21] Gonzalez [16].

can be turned into local anchors, such as cultural centres, to act as a barrier against speculation and transience. But the real and continuing crisis, he said, remains affordability: with Bronx residents paying a higher percentage of their income for rent, the city needs more affordable housing: "We know that neighbourhoods change", he said. "We're not afraid of change. What we don't want to see is those who want to remain to be displaced."[22]

This is a real battle for social justice, centered on people and on the neighbourhood community. In this context, the IPA—Institute for Public Architecture[23] is trying to define a future in which design is used as a tool for facilitating social justice and where the public sphere has a voice in all decisions that shape the built environment: they use design to challenge social and physical inequities in the city. They address urgent issues of design and policy by mobilising a network of activists, professionals, government officials and community stakeholders and seeking funding from federal, state, and municipal agencies.

The empty spaces between buildings and the unused areas of neighbourhoods are not the places where to experiment with the temporary, but to seek for permanence, stability and neighbourhood identity. Through precise design interventions, the architect Jonathan Kirschenfeld has been able to design living spaces for the homeless and underserved classes, to interrupt a process of uncertainty and temporariness by providing permanent housing.

In these projects, whose stakeholders are non-profit organisations dedicated to providing services and permanent housing to populations with special needs, unused spaces are exploited; there is a return to the typological research that characterised the studies presented by Aldo Rossi in *The Architecture of the City*, with an emphasis on the distinction between monuments and residence areas that are community spaces, places for living. There is a return to the character of the place, to the collective memory, to the sense of belonging, to all the elements able to guide the project toward solutions for reproducing the idea of the city that has been lost.

The courtyard buildings, galleries and blocks constitute types in continuity with history, they are the elements of permanence that place the project in continuity with the old and recent past, with a clear and recognisable typological structure that can transform itself and offer new possibilities for new uses.

In an article describing the work of Kirschenfeld Architects on affordable housing, the historian David Grahame Shane points out: "In a surreal twist the infill housing of the underprivileged suddenly takes center stage on the Grand Concourse to serve a new, upwardly mobile bourgeoisie, a fluid networked generation left out of New York City's current housing equation. These micro moves would invigorate the city and Grand Concourse from within and from the bottom-up, offering a new, wired, hyper-flexible Existenzminimum dwelling type to the currently endangered 'creative class' whose participation is crucial for the city's future."[24]

These micro moves wish to testify that a well-structured architecture, grounded on a formal basis that refers to the collective memory, is able to embrace the continuous transience of function and to remain vital and persistent in the urban organism.

References

1. Rossi, A.: The Architecture of the City, Oppositions Books, IAUS, NY, The MIT Press, Cambridge, Massachusetts, 1982; I ed. it. L'architettura della città, Marsilio Editori, Padua (1966)
2. Chabot, G., Villes, L.: Aperçu de Géographie Humaine. Armand Colin, Paris (1948)
3. Manzione, L.: Una reinvenzione infinita. Città, architettura, evoluzione e temporalità in Marcel Poëte, in Architettura and Tempo, Percorsi multidisciplinari di ricerca, vol. 3, p. 174. A Cura di Antonio Lavarello e Davide Servente, GUP, Genoa University Press, Genoa (2020)

[22] *Ibid.*

[23] https://www.instituteforpublicarchitecture.org/about. Last accessed 2022/08/30.

[24] Shane [17].

4. Poëte, M.: Paris, Son Évolution Créatrice. Vincent and Fréal, Paris (1938)
5. Rossi, A., I quaderni azzurri, quaderno n.1, 19 June 1968, Electa/The Getty Research Institute, Milan, 1999
6. Bishop, P., Williams, L., The Temporaray City, Routledge NY, 2012, p. 15
7. Koolhaas, R., Cronocaos, Log n.21, Anyone Corporation, NY, 2011, p.122
8. Borasi, G. (ed.): A Section of Now: Social Norms and Rituals as Sites for Architectural Intervention, p. 7. Spector Books, Leipzig, CCA-Canadian Centre for Architecture (2021)
9. Lauinger, H., Hauck, T., Frech, S., Wirth, M., Grandinetti, G., Neumann, M.: Urban Pioneers. Jovis Verlag Berlin, Senatsverwaltung für Stadtentwicklung Berlin (2007)
10. Oswalt, P., Overmeyer, K., Misselwitz, P.: Urban Catalyst. The Power of Temporary Use. Dom Publishers, Berlin (2013)
11. Oswalt, P.: Berlin Stadt ohne Form: Strategien einer anderen Architektur. Prestel, Munich (2000)
12. Oswalt, P.: Berlin, City of the 20th Century. In: Wiesel M., Berlin/Berlin, Catalogue for the Berlin Biennale, Berlin (1998)
13. Oswalt, P., Overmeyer, K., Misselwitz, P.: Urban Catalyst. The Power of Temporary Use. Dom Publishers, Berlin (2013).
14. Ascari, P., Borsari, A., Trentin, A.: International Conference. Temporary Citizenship, Architecture and City, University of Bologna (2021)
15. Schindler, S.: Model Conflicts. e-flux Architecture (2018)
16. Gonzalez, D.: Interview with Monxo Lopez. New York Times (2017)
17. Shane, D.G.: Restoring the urban dream. Affordable Housing in New York. Bauwelt **110**, 19–21 (2012)

Annalisa Trentin is an architect and full professor in the field of architectural and urban design at the Department of Architecture of the Alma Mater Studiorum—University of Bologna; since 2018 she is Coordinator of the Ph.D. course in Architecture and Design Cultures (Unibo); from 2015 to 2021 she held the position of Deputy Director of the Department of Architecture, where she has been carrying out research and interdisciplinary teaching activities since 2001. She carries out design research by participating in international workshops and competitions; she is the author of several essays on pedagogical practice in architectural design, criticism and theory of contemporary architecture.

Dissonant Times

Andrea Borsari

Abstract

The contribution aims at analyzing the different layers of meanings contained in the notion of "dissonance" from the aesthetic, cognitive and emotional point of view. To then make them available for the exploration of the different forms of "dissonant time" analyzed by contemporary thought, in particular by thinkers such as H. Blumenberg, R. Bodei, R. Koselleck, E. Traverso, E. Bloch, H. Rosa, G. Wilder and the studies on the "new urban world". In this context, notions such as the following become relevant for understanding the temporal experience of late modernity: gap between world time and life time; intertwining and overlapping of future-facing past and present; changing and at different speeds "geologically" multi-layered relations between space of experience and horizon of expectation; "presentism" and breakdown of social frameworks of memory and individualistic reification of the past; "contemporaneity of the noncontemporary"; montage and kaleidoscopic temporal dialectic; disconnection between spheres of economic, social and generational life and lack of time; multiple temporalities and cross-connections across urban spaces; concrete utopianism and multiple lines of time. So that dissonance seems to be the most appropriate conceptual and figural tool to effectively render the condition of contemporary temporal experience in its real irreconcilabilities and perspectives.

Keywords

Dissonance · Dissonant times · Resonance · Temporal multiversum and multilayers · H. Blumenberg · R. Bodei · R. Koselleck · E. Traverso · E. Bloch · H. Rosa · G. Wilder

1. The use of the musical metaphor of resonance, right from its Romantic origins as a "magic word" that gives voice to the world, has found in Hartmut Rosa's monumental work a kind of manifesto that opposes our age's tendency toward temporal acceleration, precisely the concept of resonance as a key to revealing late modernity's underlying fear for the transformation of the world and, at the same time, the driving hope for the realization of a responsive relationship with it, using the categorical tools of a sociology of our relation with the world [1]. For the preliminary purpose of delving instead into the epistemological premises of such a framework in order to discuss on another occasion the possible alternatives between the two definitions, an

A. Borsari (✉)
Department of Architecture, University of Bologna, Bologna, Italy
e-mail: a.borsari@unibo.it

A. Borsari et al. (eds.), *TEMPORARY: Citizenship, Architecture and City*, The City Project 4,
https://doi.org/10.1007/978-3-031-36667-3_12

attempt will be made here to explore briefly a complementary and opposite metaphorical direction, again of musical origin, namely that of dissonance, of its own more adequate to render the condition of temporal displacement and disarticulation of contemporary society, as sketched in the introduction to this volume and further developed in the second part of the text which follows.

Within the humanities, there has been widespread use of the term and metaphor of dissonance over the past three decades. For example, following the canonical reference to Turnbridge and Ashworth [2], the dissonance implied in "Dissonant Heritage" involves as much an idea of discrepancy and incongruity, i.e. a discordance and lack of agreement and consistency, as it does an allusion to the field of music and harmony theory with its classification of disharmonic combinations of sounds, which in turn recalls the psychological notion of cognitive dissonance as a state of psychic tension caused by the simultaneously holding of mutually inconsistent attitudes. Therefore, in parallel with the task of developing an analysis of the three main sources of dissonance (implicit in commodification; implicit in place products; implicit in the content of the message), it seems appropriate to reflect on the ways of knowing and the criteria of relevance that act in the definition of what is dissonant, starting from their aesthetic dimension. To this end, this contribution aims to analyze the different layers of meaning that the term sediments in different areas: musical, literary and artistic in what was defined at the beginning of the last century as "emancipation of dissonance", in cognitive psychology according to its implications of "pressure to reduction" and in social psychology as "emotional dissonance" (§ 2). Once the set of meanings that converge in the notion of dissonance has been articulated, and after showing its possible implications in parallel in memory studies (§ 3), the second part of this paper turns to the exploration of the possible ways in which divergent, juxtaposed and multilayered temporal dyscrasias—finally, dissonant times—can be detected, connected and coexist in contemporary experience of time. For how they can be analyzed by drawing on the elaborations of Hans Blumenberg and Remo Bodei (§ 4), Reinhart Koselleck and Enzo Traverso (§ 5), Ernst Bloch and Hartmut Rosa (§ 6), field studies of new urban worlds and political philosophical approach to concrete utopianism (§ 7).

2. If we understand "epistemology" at its broadest level of determination, as a second-level investigation, i.e. a reflection on the very modes of knowledge that are adopted in a given field, in our case, it is a question of analyzing the different layers of meaning that the reference to the musical metaphor of "dissonance" produces. In particular, the levels of meaning related to the following should be made explicit: the musical sense of "musical dissonance"; the wider aesthetic sense in art and literature; the so-called "emancipation of dissonance"; the sense more closely linked to cognitive processes, of "cognitive dissonance", and the psychological sense of "emotional dissonance".

Layer of meaning: Musical Dissonance. From the Latin *dissonantia* (Boethius, *De Institutione musica*) is etymologically "the fact of not sounding with" and is therefore defined in relation to consonance: it is non-consonance. The term indicates two or more sounds (forming an interval, a chord or a harmony) that seem heterogeneous, and whose simultaneity requires resolution, a consonance. But—suggests Lyse Florenne –the boundaries between consonance and dissonance are variable and bring into play mathematics and physics, physiology and psychology, as well as customs and traditions. Particularly in the Renaissance and Baroque music, it was used for expressive purposes of tension, torment or restlessness, of waiting for the resolution and restoration of consonance. Nevertheless, the imbalance, heterogeneity, strangeness and extraneousness of dissonance often constitute an exciting pleasure for the ear and spirit [3].

Layer of meaning: Aesthetic Dissonance. In a figurative sense, dissonance has come to indicate a disruptive element in other artistic forms: in painting, dissonance is used to refer to colors that do not harmonize well with each other and about which we can say that they are astonishing; in literature, dissonance of style is used to denote a heterogeneous mixture of, for example, the noble and the vulgar, the sad and the laughable; a dissonance in the discourse can be a break in style that destroys the harmony and can have an exciting stimulating effect or a hurtful shock.

Layer of meaning: Emancipation of Dissonance. The year 1910 constitutes, according to Tom Harrison's reconstruction, the turning point in the evaluation of dissonance and finds expression in the formula adopted by Arnold Schönberg in his *Theory of Harmony* (1911): the emancipation of dissonance. Its novelty consists in the manifestation of a torment, in the transcription of a dilemma. Similar assertions, in the years when Schönberg abandoned classical harmonic structures for atonality, are supported by other artists, for example, Vasily Kandinsky. According to him, the harmony of the time can only be that of "contrasts and contradictions": "struggle of tones, loss of balance, fall of principles"... "seemingly purposeless tensions, impetuses and nostalgia: this is our harmony" [4]. Following Schönberg, as the ear becomes acclimatized to a sonority within a particular context, the sonority will gradually become 'emancipated' from that context and seek a new one. The emancipation of the dominant-quality dissonances has followed this pattern, with the dominant seventh developing in status from a contrapuntal note in the sixteenth century to a quasi-consonant harmonic note in the early nineteenth. By the later nineteenth century, the higher numbered dominant-quality dissonances had also achieved harmonic status, with resolution delayed or omitted completely. The greater autonomy of the dominant-quality dissonance contributed significantly to the weakening of traditional tonal function within a purely diatonic context. Following Wilhelm Worringer, "The imaginative life of mankind obeys a very simple law; it lives on antithesis" [4]. No un-historical generalization: but it results from a specific historical moment. Emancipating dissonance is not just about recognizing, suffering or reflecting that dissonance, it is about making it the basis for a new kind of art by virtue of the metamorphosis and dislocation of form. The pictorial motifs are transformed into the cipher of a network of universal forces in which unions and contrasts are one and the same. Schönberg later abandoned atonal music and Kandinsky his transitional style: aspiring to be prophets of a more sophisticated consonance than language could comprehend, these artists became witnesses and martyrs to the dissonance they themselves had liberated. Finally, what appears to be emancipatory is the conflict between consonance and dissonance through which the scenario, the larger cultural scenario of dissonance that these artists produce, expands. Dissonance is not what is emancipated in the emancipation of dissonance. Dissonance itself—the search for resolutions not found—bears the emancipatory charge. So that we can understand it as an emancipation of dissonance, considering the genitive "of dissonance" as a subjective and not objective one [5].

Layer of meaning: Cognitive Dissonance. Concept introduced to describe the condition of individuals whose beliefs, notions, opinions conflict with each other, or with trends in behavior or with the environment in which the individual operates [6]. It may arise from logical inconsistency between two incompatible opinions, from cultural customs that are at odds with the context in which one operates, from inclusion when a given opinion is included in a more general opinion (e.g. someone votes left but is right-wing or vice-versa: discordance between corresponding cognitive elements), from past experience that has created beliefs that do not fit in with new experiences. The presence of dissonance gives rise to pressure to reduce or eliminate it, prompts action in order to obtain reduction through change of opinion, behavior or environment, and addition of new consonant cognitive element. It cannot always be reduced [7].

Layer of meaning: Emotional Dissonance. By emotional dissonance the psychology means the discrepancy between expressed and felt emotions, when the former are in accordance with emotional norms but are discrepant with the emotions felt 'inwardly' and which one would like to express. Emotional deviance is the behavioral expression of emotional dissonance, genuine acting is the behavioral expression of emotional consonance. When a condition of emotional dissonance or deviance is present, emotional labor and elaboration represent the effort that is made to try to reduce this discrepancy. It requires to focus on criteria to which we culturally refer to judge whether an emotion is in consonance or dissonance with the emotional rules of the context [8].

3. The relationship between consonance and dissonance describes a condition similar to the insurmountable tension between remembering and forgetting, as far as it relates to the internal dissents within the sphere of public or collective memory and the field of cultural heritage, from whose example we took our cue. Some possible directions for this research include three perspectives that diverge from the unilateral results of excessive remembering, the hypermnesia that crowds memories and, by hardening them, makes them indiscernible, and of excessive forgetting, the amnesia that reacts to an excess of voluntary remembering but ends up confusing the physiological processing of forgetting with the questioning of immovable history underpinned by the ethical link with memory. The three examples are those of providing spatial experience through the building of places having a strategy of reawakening "dormant memories", the conflicting relationship between monuments as aids to memory and counter-monuments as attempts to escape the paradoxical erasure of memory induced by the former, and finally of the recovery of a perspective derived from Georges Perec's infra-ordinary to experience crucial places of remembrance where time and human destruction have left nothing but pale traces of the horror that took place in them, as in the exemplary case of Auschwitz Birkenau [9]. In relation to dissonance and to dissonant times, a consequence can be drawn from this comparison according to which dissonance can be a way of giving voice to differences, of seeking a terrain in which opposing terms without overcoming each other in a higher chord become a form of antinomian coexistence, a tension that, instead of resulting in conflict or harmonize a priori in a conciliatory perspective, keeps different forms of solution open.
4. Once we have gained a definition of the notion of dissonance, it is a matter of comparing it, again in terms of sketching out a research track, with the articulations of modern temporality and the possibility of reading them as dissonant times. It dates back to 1986 when Hans Blumenberg attempted to give shape to the basic contrast that opposes life time and world time, a sign of the temporal dyscrasia that afflicts modernity and undermines the structures of consciousness that have governed the forms of experience accredited up to that point [10]. "Ars longa, vita brevis": as long as the state of the world appears to be immobile or seems to be transforming only slowly, bound to repeat itself by strict obligations and institutions, the ancient adage of wisdom appears to resonate as reliable and true, while the feeling that life goes on after us cannot present itself as a feeling of having had to give up something essential. In this perspective, the extreme resentment of a sense of deprivation arises from the perception of reduced possibilities as a consciousness of a backlog of missed life possibilities, based on the additional assumption that the realities given as the conditions of one's life might have been so different. Any divergence between the time of the world and the time of life potentially contains the scandal, that is the basis—as Blumenberg reminds us—of every apocalyptic thrust, which for the individual is the fact that the world continues to exist unperturbed beyond the end of his or her life. It is not about heroic self-enjoyment or aesthetic distance, i.e., it is not about the stoic *vir impavidus*, who endures amidst the hail of the world's rubble, nor is it about the spectator of the shipwreck to whom the epicurean Lucretius makes enjoy his foreignness. Instead, it is

> a matter of having to endure the indifference of the world, of its existence before or after us, as *the* rejection of meaning. Hence comes the desire that, given our transience and finiteness, everything else must also be transient and finite. The desire that time of the world and time of life end up coinciding, which has found the extreme historical case in the violent reduction of one to the other emblematized according to Blumenberg by the personal and public history of Adolf Hitler, where the psychic disposition is not to accept and wait, to beg for acceleration and to keep oneself out of the world, but to *pursue* the end [11].

In that case, the borderline case of paranoia, the one and only life a man possesses becomes the condition for a historical and political empowerment to take place. Of the failure of the purpose of his own life he can then make the failure of the meaning of the world: if he is doomed to perish, whatever factual circumstances have compromised his life program, then everything is doomed to be at an end. A single existence defines to itself its own meaning precisely by claiming to be something after which there must be nothing else and, broadening Freud's language, one could speak of "absolute narcissism". As Hitler himself had occasion to tell his aide for the Luftwaffe, "We do not capitulate, ever. We may perish, but we will drag a world with us" [12]. In this, for its consequences, terrible decision to synchronize time of life and time of the world is defined the destruction of the essential institutionality of historical time, since institutions are based precisely on the fact that the time of life is not the measure of all things, that on the contrary it is necessary to make arrangements that override the limits of life, to fix and accept traditions that go beyond it. In order to be able to endure this one must be able to relativize one's life, in other words, one must possess the horizon of the world. If, again according to Blumenberg, epochs and cultures represent second-degree life worlds in themselves, their structures of meaning give rise to what—thanks to and within their horizon, that is, by contemporaries and co-participants in their conferral of meaning—is considered the obvious. The emergence of contingency, in the opening of the temporal gap—the *Zeitschere* or time scissors—between the time of the world and the time of life, inevitably raises the problem of accepting this incongruity. But there is a viable way to escape the temptation to reconstitute both to congruence. It is the way Blumenberg describes as decoupling the time of life from the time of the world, beginning with the possibility of avoiding the trauma of their disproportionality by the descriptive affirmation of their incomparability. If leaving the life world means that life time and world time begin to diverge, asking for meaning is already a phenomenon of deficiency, whose therapies never restore the original state but, if anything, its surrogates. However, it is possible, in agreement with Blumenberg, to enact de-dramatizations, decouplings or mitigations (*Entschärfungen*) that, by disengaging the time of life, allow the regaining of the time of the world. Such de-dramatizations have occurred historically. For example, as idealization of space and time that made them the conditions of every possible experience. Or as the genesis of time in consciousness for phenomenology in which space and time are forms by which consciousness adapts to its own narrowness, its own disproportionality to the world. Or, from the point of view of the later Husserl, as non-indifference of what is because it has been, whereby leaving memories, becoming remembrance, would be precisely the sense of a world that has arisen exclusively for the purpose of putting intersubjectivity into operation, and according to which decisive determination of our consciousness is the fact that we can say that I remember not only me but also these and these others. According to this renewed conception of anamnesis, the concept of recollection must be broad enough to encompass the totality of a world that is past, it must be the memory of a world [13].

As Remo Bodei reminds us, bringing these considerations of temporal dyscrasias back to a phenomenology of the present, individual consciousness today retains less of the past and is losing the ability to unwind it to rejoin the present. Human memory, in fact, is constituted by the permanence in us of a past considered significant, the result of a selection, conscious or unconscious, of events and notions that mark the different stages of life. This is not merely a passive recording of events, but a quest, a Proustian *recherche,* directed toward the future through an endless network of references and allusions. If the mutual isolation of present, past and future exceeds a certain threshold, the symptoms of a true pathology of time manifest themselves, and events deprived of any intrinsic order of succession or coexistence are placed in parameters that are unclassifiable according to the usual criteria of experience [14].

5. Returning in his later years to the relationship between space of experience and horizon of expectation, more than two decades after *Future Past* (1979), Reinhart Koselleck insisted on the multilayered dimension of historical temporality. In his *Zeitschichten* [Time layers] (2000) he questions the two categories, noting how historical time is notoriously difficult to make intuitable and it is necessary for it to resort to spatial background meanings that can only be described metaphorically. They are, precisely, the two anthropological categories of space of experience and horizon of expectation. No historical action can be accomplished without relying on the experience and expectation of the actors involved in it. Both categories are available to thematize historical time, since past and future are delimited in the present of experience and expectation, and even in empirical terms they orient the individual units of action in political and social movements. The two temporal extensions condition each other in varied ways, as historical knowledge is gathered in experience and transposed not seamlessly into expectation. If this were possible, however, history would only, fundamentally, repeat itself. Similarly, as remembrance and hope, the two dimensions have a very differentiated status. What is expected for the future is manifestly circumscribed in other ways than what has been experienced in the past. Expectations to which one aspires are exceeded, experiences made are instead gathered. Space of experience and horizon of expectation should therefore not be put in statistical relation to each other, but constitute a temporal difference in the now, as they delimit and establish a relationship between past and future in a different way each time. Koselleck's historical thesis is that, since the modern age, the difference between experience and expectation has steadily increased. Or, more precisely, that the modern age has been conceived of as a new time in which expectations have become increasingly distant from the experiences that have been made up to that point. Since the eighteenth century, with a radical turn marked by the Enlightenment, this process has profoundly altered the entire political and social vocabulary, which has adopted an internal temporal structure in which the experience share and the expectation share have completely changed their respective weights. Concepts almost deprived of experience content and with increasing claim to expectation are asserted, defining a temporal asymmetry that makes them more properly anticipations. We are facing a temporal tension that, once it arose, also characterizes our political and social systems today. Political and social concepts thus become guiding devices of the changing historical movement. They do not merely record or survey existing states of affairs, but become factors in the construction of consciousness and orientation of action [15].

The consequences Koselleck draws for the definition of historical temporality in terms of social history are related, in addition to the experience-expectation pair, to the temporalization of modern history and the ability to separate and distinguish between different temporal planes. In the field of spatial metaphorics, "layers of time" refer to geological formations that penetrate with different amplitude and different depth and that in the course of the so-called history of the earth have changed with different speeds and differently spaced. Its translation into human, political, or social history and the history of structures makes it possible to separate analytically different temporal planes or levels, on which people move, events take place, or their long-term preconditions are investigated. History always has to do with time, with times that not only metaphorically but also empirically remain related to spatial specifics. "History", in German *Geschichte*, refers to the verb *geschehen*, to happen, which literally means to keep moving, to move forward. Every historical space is constituted by force of time, through which it is measured, in order to be dominated in economic or political terms. But the spatial metaphor that pluralizes the concept of time has a specific advantage. As in the geological image, the layers of time refer to temporal planes of different durations and origins that are nevertheless present and effective at the same time. Even the contemporaneity of the noncontemporary (*Gleichzetigkeit des Ungleichzeitigen*), one of the most revealing historical phenomena, is brought back, thanks to the "time layers," to an overall concept. All conflicts, compromises and consensus building can be traced back, in terms of time theory, to tensions and breaklines, and can be triggered by them. This plural dimension also allows us to connect to metahistorical factors that are not available to humans, such as geographical and climatic conditions that can be influenced by them but which they cannot control and direct [16].

In the wake of Koselleck's elaborations, the historian Enzo Traverso has come to define the contemporary relationship with temporality under the banner of compression in the present, thanks to the notion of "presentism." With this expression, coined by François Hartog, he defines the regime of historicity generated by neoliberalism as the perception and representation of time crushed and compressed in the present: the past no longer announces the future, no longer contains any promise of redemption, past and future remain encapsulated in an eternal present. The other salient feature of the presentist historicity regime—Traverso continues—lies in its apolitical nature: "The new forms of capitalism progressively erase the social frameworks of memory, destroy their traditional channels of transmission and deprive collective action of any historical legacy." [17]. This process produces a displacement of memory into the individual sphere, however much it is expressed through a myriad of social vectors ranging from commemorations to the culture industry, and it ends up being reduced to a set of stories, images and emotions that replace collective reflection and action. From such depoliticization also descends–according to Traverso—a further characteristic of presentism, the "reification of the past": "The end of transmissible experience generates the 'places of memory,' a set of sites, objects, images and symbols that organize the past as a kind of heritage, an inherited property, fit only to be preserved in a museum, ready for commemorations, to be turned into merchandising for the cultural industry and consumed individually" [18]. In the presentist iron cage, the past no longer generates utopian imaginations because its perception is structured by mercantile consumption. Just as it privatizes the imaginary, the future as a project of individual success, neoliberalism tends to privatize the past, making the self its observatory and at the same time its laboratory. Whereas, to force the iron cage, one must make the effort to inscribe individual destinies in a larger historical scenario.

6. To the plural co-presence of different times, their possible differentiated dilation and the detection of the utopian latency of the present had been devoted a large part of Ernst Bloch's work, starting, precisely, from the notion of the non-contemporaneity of the contemporary. He focused on defining the issue of spatial and temporal discrepancies, that is, the fact that no matter how much we live in the same chronological time and on the surface of the same planet, time and space are not the same for everyone. To denote the mismatch between non-congruent historical times coexisting in the same chronological present, Bloch uses the term "Ungleichzeitigkeit," literally "non-contemporaneity" or,

> affirmatively, "anachronism." Through it, Bloch obtained the theoretical key to the interpretation of Nazism and the structure of social classes in Germany, but also, more broadly, to understand how different historical times are condensed and intertwined in the current time and we are dispersed and distributed along different temporal and spatial axes, along gradients that are both historical and geographical. The opposition between big and small town, city and country, center and periphery are also temporal, historical. The small town and the countryside, with their inhabitants, are not simply backward compared to the metropolis, they do not simply live in a "gap time," a time of backwardness, but in another time, which does not coincide with that of big cities and other social groups [19].

Bloch proposes to consider time in a plural and multiverse key, as in the case of dream time, which does not coincide with the quarter-hour in which one dreams. The instant, invisible and without depth, is outside time and touches with eternity. We have "immense moments" that are the revelation of something, and if we mean by eternity not a long time but the fullness of living, then eternity is not necessarily an infinite time but is found in those moments of great intensity of experience, grasping eternity in the moment. It is an idea of non-homogeneous time in which each minute is not the same as the next and of the fullness of experience. The possible connection between temporal gradients is not implemented for Bloch through a kind of harmonization, but through an enhancement of the partial and divergent aspects, taking as a model the experiences of the artistic avant-gardes of the early twentieth century, particularly Expressionism and the technique of montage. The Expressionists' gaze shines in a new, and thereby instantly exploded, contemporary depth, and produces a distortion of the flatness of reality aimed at revealing the real violence and lacerations that humans and things suffer in this world. Similarly, montage blows up the connections considered "natural", the parts no longer coincide, they are removable and reassembled, to the point of the possibility of constructing—in literature—a coexistence of times based on the kaleidoscope model. In the more properly musical sphere, too, Bloch points to a multi-layered dialectic through polyrhythmics and counterpoint of different tempos, aimed at a critical rather than contemplative totality [20].

On the sociological level, Hartmut Rosa (2010) defined the ways in which temporal acceleration tends to increase the degree of estrangement and alienation. On the societal level, different degrees of acceleration are possible for different social spheres. While the speed of economic transactions and technological innovations seems to have grown exponentially in recent decades, the pace of politics has not accelerated at all, Rosa notes, and the pace of cultural reproduction, i.e., the generational passage of symbolic knowledge, has accelerated very little. The same process that makes social, cultural and economic changes accelerate slows down the process of forming democratic political will, and this leads to a desynchronization between politics, life and socioeconomic evolution. The dysfunctional desynchronizations of the late modern time fabric include those internal to economic life, such as the mismatch between the timing of financial markets and the timing of the real economy, those internal to the world of life, which is being dynamized to such an extent that there is very little intergenerational stability and generations are virtually living in different worlds, threatening a collapse of the symbolic reproduction of society. Moreover, society's creative ability to provide truly innovative responses to new conditions seems to require a considerable amount of free or otherwise abundant time, which for the bulk of the population is completely unavailable. So that it would be precisely the late modern society's tireless struggle to change and be incessantly in motion that would undermine its ability to truly innovate and adapt in a creative way. The hyperdynamic surface of such societies would thus hint at a very solid form of sclerotization and freezing [21].

7. On the more decidedly empirical level of the analysis of new urban worlds, authors who have devoted themselves to field analyses, especially in Africa and South Asia but with implications for Latin America as well, such as AbdouMaliq Simone and Edgar Pieterse, have also mapped a series of multiple temporalities and transversals of connection that operate across and through urban spaces [22]. They noted how within the "real urban " it is difficult to draw clear demarcations between functions, flows, territories and events. Similarly, institutional life has been largely distributed across a network of seemingly divergent practices and actors, prioritizing the non-contractual and the disavowal of overly congealed forms of inhabitation. The very fact, noted by Ravi Sundaram (2015) in the wake of Arjun Appadurai's insights into the different "scapes" of the migrant experience, that now everyone, no matter how settled they may be, possesses and lives through various media devices, making things public in ways that cannot be controlled, as individuals circulate through various experiences and information whose dispositions are difficult to anticipate [23]. Any attempt to cope with such a volume of interactions through data collections and simulations runs into the perverse effect, pointed out by Luciana Parisi, whereby an increasing volume of data is generated that cannot be compressed and, thereby, readily systematized. She stresses that data are always mobile and are always being recombined and retrieved, moved around, stored and restored in different settings, which are also mobile. So, while these "histories" of data cannot always be computed, they can still affect any algorithmic operation that attempts to interrelate the variables at work in the production of an "actual project": ironically the actuality of any project then has to consider its own incomputability [24].

For their part, Simone and Pieterse, propose a strategy based on several moves, among which re-description, the detection of paradoxical situations and the possibility of inhabiting divergences and dissonances play a central role. The notion of "re-description" serves to interrogate how specific spaces, built environments and ways of living according to different times can be re-described not as elements that make evident certain principles, macro-forces or structural articulations but as aspects of what they *might* be, as other dimensions of space or states of existence. This is not simply a matter of imagination, fantasy or forward visualization, they suggest. Instead, the task at hand is—with motions that would be of detecting the potential elements of a concrete utopia, although not deliberately here—to re-describe existing conditions as components of a process that could take place even now but is opaquely occluded or rendered inoperative. That is, it is a matter of grasping an ambiguity and ambivalence of what is in front of us and that involves simultaneously a sense of aspiration, of making things different, but also of seeing in what exists something other than what we think we are seeing. Not dissimilarly, if the analysis of flows and fluxes suggests that precariousness and uncertainty represent a material and psychic threat, they may also provide an opportunity to reimagine and re-enchant the world: "In other words, we cannot access an imaginary about alternative urban futures without confronting the violent impossibility of that future" [25]. Of great importance is to maintain a multiplicity of story lines; divergences will have to be part of the same story, a story of density within the relationship between all things, even as a unique urban future. The possibility of coping with the very blockage of resonance that even in some detected aspects of the most unlikely urban realities seems possible, therefore, appears to depend, in this perspective, on the ability to inhabit dissonant times and account for the divergent stories connected to spaces and temporal processes that are indeed divergent [26].

Political philosopher Gary Wilder is also approaching an appreciation in a concrete utopian key of the plurality of co-existing times. In his recent (2022) prospection on the political relationship between solidarity and temporality, he discusses precisely the possibility of defining concrete utopia as an attempt to identify "possibilities for alternative arrangements that may already dwell within, or be emerging from, the nonidentical order that actually exists" [27]. Such an attitude requires analysis, imagination, and action not constrained by a realist assumption about "chronological sequence, self-identical periods, and categorically distinct tenses": "It requires a political orientation that does not shackle historical temporality to the logic of abstract clock-time and the tyranny of tenses. [...] This requires attention to the time as a real abstraction—to the typical velocities, rhythms and scales through which lives are lived, to the specific ways that the relation between past, present and future is figured and to the temporal dimensions of various social institutions, ideologies and practices. It requires attention to how certain kinds of social arrangements presuppose and create historically specific kinds of time (and vice versa). It also means attending to the multitemporal and untimely dimensions of social life" [28]. In similar form, analysis, imagination and action of concrete utopia intoned to an untimeliness require us to reflect on our temporal implication in relation to our historical present. Wilder thus proposes to treasure, in addition to the Blochian lesson on noncontemporaneity, various theories on the multiplicity of timelines. Such as Jacques Rancière's suggestions on anachronism, which bases the meaning of time and the condition for acting historically on the "multiplicity of lines of temporality that are present in any 'one' time"; Siegfried Kracauer's "cataract of time", Mikhail Bakhtin's "essential multitemporality" of modern life, and—again—Koselleck's claim that the "new time" of modernity is a unity composed of a "coexisting plurality of times" [29].

We have with this come to the threshold of the possible comparison between resonance and dissonance strategies announced as the development of the research in the opening. Thus, dissonance is connoted: aesthetically as discrepancy and incongruity, lack of agreement and coherence, rejection of harmony, representation of tension, torment and restlessness, disruptive element in form, mixing of the heterogeneous and contrast of tones and colors, "emancipated" presence of antithetical terms; cognitively as a conflict between knowledge and actions, and, emotionally, as discrepancy between prescribed and aspired emotions. On the other hand, dissonant times present the salient features: of showing the gap between world time and life time to the extreme of loss of meaning and deinstitutionalization of time; of possible intertwining and overlapping of future-facing past and present; of changing and at different speeds "geologically" multi-layered relations between space of experience and horizon of expectation; of "presentism" and risk of one-dimensional compression of time with the breakdown of social frameworks of memory and individualistic reification of the past; of paradoxical contemporaneity of the noncontemporary as the co-presence in the same present space of different times and forms of life, open to both utopian or retrograde latencies and different qualitative experiences of the same time; of disconnection and assemblage between the parts as a composite and kaleidoscopic temporal dialectic; of disconnection between spheres of economic, social and generational life and lack of time to unlock the only seemingly hyperdynamic societies of late modernity; finally, of multiple temporalities and cross-connections across urban spaces, as well as, of plurality of times in a political perspective of concrete utopianism. So that dissonance seems—at present—the most appropriate conceptual and figural tool to effectively render the condition of contemporary temporal experience in its real irreconcilabilities and, perhaps more than resonance, also to allow in perspective to hold together its harshness and the possibilities of reopening.

References

1. Rosa, H.: Resonance. A Sociology of Our Relationship to the World, Polity, Cambridge, (2019). See also: Sensibilia 13 Resonance, Studi di Estetica, 17: https://journals.mimesisedizioni.it/index.php/studi-di-estetica/issue/view/98 (2020)
2. Turnbridge, J.E., Ashworth, G.J.: Dissonant Heritage. The Management of the Past as a Resource in Conflict, Wiley, Chichester (1996)

3. Florenne, L.: Dissonance. In : Souriau, E. (ed.) Vocabulaire d'Esthétique, p. 594. Puf, Paris (1990). Where not referred to a printed translation, all quotations from languages other than English in this text should be understood as translated by the author of this contribution.
4. Harrison, Th.: (1910) The Emancipation of Dissonance, University of California Press, Berkeley-Los Angeles-London (1996)
5. Harrison, Th.: 1910. The Emancipation of Dissonance, p. 221 (1996)
6. Festinger, L.: A Theory of Cognitive Dissonance, Stanford University Press, Stanford (1957). See also: Cooper, J.: Cognitive Dissonance. Fifty Years of a Classic Theory, Sage Publications, Los Angeles (2007)
7. Festinger, L.: A Theory of Cognitive Dissonance, p. 17 (1957)
8. Rafaeli, A., Sutton, J.: Expression of emotion as part of the work role. Acad. Manag. Rev. 12 (1), 23–37 (1987). https://doi.org/10.5465/amr.1987.4306444
9. For an early development of this subject, see: Borsari, A., Leoni, G.: Hypermnesia and amnesia: remembering (with) the body and post-conflict memorials and architectures. Heritage Memory Conflict J. 2, 29–38 (2022)
10. Blumenberg, H.: Lebenszeit und Weltzeit [Lifetime and world time]. Suhrkamp, Frankfurt am Main (1986)
11. Blumenberg, H.: Die Kongruenz von Lebenszeit und Weltzeit als Wahn [The congruence of life time and world time as madness], in: Lebenszeit und Weltzeit, pp. 80–85
12. Blumenberg, H.: Lebenszeit und Weltzeit, p. 80
13. Blumenberg, H.: Entschärfungen: Abkopplung der Lebenszeit—Zurückholung der Weltzeit, [Dedramatizations: Decoupling of the life time—Retrieval of the world time], in: Lebenszeit und Weltzeit, pp. 86–98
14. Bodei, R.: Dominio e sottomissione. Schiavi, animali, macchine, Intelligenza Artificiale [Domination and Subjugation. Slaves, Animals, Machines, Artificial Intelligence.], Il Mulino, Bologna, pp. 376–379 (2019)
15. Koselleck, R.: Zeitschichten. Studien zur Historik [Time layers. Studies in history.], pp. 331–335. Suhrkamp, Frankfurt am Main (2000)
16. Koselleck, R.: Zeitschichten, pp. 9–16
17. Traverso, E.: Presentismo, in: La tirannide dell'io. Scrivere il passato in prima persona [Presentism, in: The tyranny of the self. Writing the past in the first person], Laterza, Roma-Bari (2022), pp. 143–167, here p. 153. For references to Koselleck, see pp. 147, 160, 166
18. Traverso, E.: Presentismo, p. 153
19. Bloch, E.: Ungleichzeitigkeit und Pflicht zu ihrer Dialektik (1932) [Non-contemporaneity and duty to its dialectics], in: Erbschaft dieser Zeit [Inheritance of this time], Gesamtausgabe, vol. 4, pp. 104–160, Suhrkamp, Frankfurt am Main (1962). The reconstruction of Bloch's philosophy suggested here follows that developed by R. Bodei, Multiversum. Tempo e storia in Ernst Bloch [Multiversum. Time and history in Ernst Bloch], Bibliopolis, Napoli, (1979)
20. Bloch, E.: Geist der Utopie [Spirit of Utopia] (second version, 1923), Gesamtausgabe, vol. 3, pp. 100–108. Suhrkamp, Frankfurt am Main, see Sprechgesang, synkopischer Rhythmus und akkordische Polyphonie (1964)
21. Rosa, H.: Alienation and Acceleration: Towards a Critical Theory of Late Modernity, NSU Press, Natchitoches-LA, on pathologies of desynchronization, see chap. 11 (2010)
22. Simone, A., Pieterse, E.: New Urban Worlds. Inhabiting Dissonant Times, Polity Press, Cambridge (2017)
23. Sundaram, R.: Publicity, transparency and the circulation engines. Curr. Anthropol. 56, 297–305 (2015)
24. Parisi, L.: Contagious architecture: computations. Aesthetics and Space, MIT Press, Cambridge MA (2013)
25. Simone, A., Pieterse, E.: New Urban Worlds, p. 31
26. See Simone, A., Pieterse, E.: A Story About Stories, in: New Urban Worlds, pp. 183–198
27. Wilder, G.: Concrete utopianism. The Politics of Temporality and Solidarity, p. 9, Fordham University Press, New York (2022)
28. Wilder, G.: Concrete Utopianism, p. 155
29. Wilder, G.: Concrete utopianism, p. 150, on multitemporality, see the two first chapter of part II, pp. 139–190

Andrea Borsari teaches Aesthetics as a full professor at the Department of Architecture of the University of Bologna, directs the series "La vita e le forme" (Bup) and is deputy director of "Iride. Rivista di filosofia e discussione pubblica" (Il Mulino). He was recently visiting professor at UCLA, ENSA-Paris La Villette, the 4 Cities Unica Euromaster in Urban Studies—University of Copenhagen and the University of Sichuan, and visiting fellow at University of Amsterdam, University of Buenos Aires and Brown University—Providence. Among his publications: *Mondo, cose, immagini. Sulle forme dell'esperienza estetica* (Bup, 2018), *Mimicry. Estetica del divenire animale* (Mimesis, 2018) and, as editor, *Philosophical Anthropology and Critical Theory* (Iride, 02/2022), *Aesthetics of the Anthropocene* (CPCL, 1–2/2022, with P. Ascari).

For a Lemmary of Temporary Citizenship

Preface During the course of the research pathway we undertook, which led to the conference documented in the previous part of this book, the need emerged to investigate the notion of citizenship in an additional or perhaps preliminary dimension to the realms of its formal definition. To such a definition, insofar as it is complemented by the recognition of instances and rights that have not yet been integrated into the legal systems, it seemed to us that the outcomes of the analyses concerning performative citizenship could also be traced. A different approach, then, was provided to us by the operative hypothesis of posing the problem of how practices and usages can force or deconstruct the statute of citizenship a moment before their recognition or failure, sticking to the historical and conceptual plane of an effective citizenship (because it is still internal to what happens in the city) but not necessarily codified. The name we thought it plausible to give to this plan is that of *temporary citizenship*. And it is within this framework that we then asked the doctoral students in Architecture and Design Culture of the Department of Architecture at the University of Bologna (XXXVI Ph.D. cycle) to express what the implications of citizenship might be in the perspective of their individual research, accounting for them in the construction of the lemmas that you will be able to consult in the next pages.

Assemblages/Decentralised

Andrea Cattabriga

Abstract

The concept of citizenship has traditionally been associated with privilege and the rules defined by nation-states. However, the increasing integration of digital technology calls for a reconsideration of the concept of citizenship as a system of relations rather than a static condition. The assemblage theory offers a way to understand citizenship as a dynamic, temporal, and spatial concept involving continuous transformation and networks of interdependent relationships. This shift from a fixed understanding of belonging to a more fluid and dynamic concept requires a reevaluation of the role of non-human actors and the importance of multiple, shifting narratives of time. Emerging digital technologies are enabling new forms of citizenship that are no longer tied to territorial jurisdictions, but instead are based on participation in distributed virtual communities. The development of decentralised technologies such as blockchain offers the potential for creating new institutions that enable a more agile and rapid recodification of citizenship status. However, the decentralisation of citizenship also raises concerns about the potential for centralised control and the need to protect decentralised activities. A key challenge is to imagine a strategy that can combine centralisation and decentralisation in a way that enables new forms of citizenship based on belonging to extraterritorial, distributed, and decentralised values and affinities.

Keywords

Assemblages · Digital technologies · Decentralised citizenship

1 Assemblages

The concept of citizenship has often been associated with a status of privilege, thus corresponding to the rules that nation-states define as prerequisites for being a citizen, placing the concept of belonging to a political community as a foundational element of citizenship [1]. Especially when enriched with a 'temporary' qualification, it is perhaps useful to define citizenship as a system of relations rather than a condition.

Today's deep integration with the digital makes it even more urgent to qualify the phenomenon from the perspective of the interrelation

A. Cattabriga (✉)
Department of Architecture, University of Bologna, Bologna, Italy
e-mail: andrea.cattabriga@unibo.it

A. Borsari et al. (eds.), *TEMPORARY: Citizenship, Architecture and City*, The City Project 4,
https://doi.org/10.1007/978-3-031-36667-3_13

between actors of various natures: biological, computational and material.

Deleuze and Guattari [2] define the term assemblage as the way in which material contents (bodies, actions, passions) and utterances (statements, plans, laws), are connected.

The set of relationships that connect an assembly must be imagined as a network rather than a structure, in the sense that the parts retain their own independent identity outside these relationships, which in turn are produced by a multiplicity of forces that determine 'rhizomatic', non-linear relationships (ibid.).

The assemblage theory thus offers a reference to define temporary citizenship as a system of relations rather than a status defined by static properties, emphasising both the temporality and spatiality of this dynamic [3]. These kind of arrangements are associated with continuous actions of transformation in a realm of heterogeneous components that co-function [4], creating networks of interdependent relations with the local environment and the world at large. This approach requires a shift from the idea of belonging as a fixed state to a more fluid and dynamic understanding of citizenship.

As stated by Müller and Schurr [5], "assemblage is a mode of ordering heterogeneous entities so that they work together for a certain time". A condition—which questions the very concept of belonging and bonding. Collier and Aihwa [6] identify what they call global forms, as phenomena distinguished by their "capacity for decontextualization and recontextualization, abstractability and movement, across diverse social and cultural situations and spheres of life". When territorialized in assemblages, "they define new material, collective, and discursive relationships":

> In relationship to 'the global', the assemblage is not a 'locality' to which broader forces are counterposed. Nor is it the structural effect of such forces. An assemblage is the product of multiple determinations that are not reducible to a single logic. The temporality of an assemblage is emergent. It does not always involve new forms, but forms that are shifting, in formation, or at stake. As a composite concept, the term 'global assemblage' suggests inherent tensions: global implies broadly encompassing, seamless and mobile; assemblage implies heterogeneous, contingent, unstable, partial and situated.[1]

Moreover, on the dimension of defined and unambiguous time (adherence to a single flow-narrative generated by the authority that constructs the rhetorical narrative and recognises the citizen's status within it), it is perhaps possible to focus on an essential component of temporary citizenship. Indeed, it is not to be understood as related to a cognition of the 'temporary' span of a single possible timeline, but rather to a time track belonging to a different narrative, to a multiple conception of time itself [1]. It is the time of assemblages, those forms in continuous transformation understood by Delanda [7] as the new conceptual foundation of society. A perspective in which change takes place through collective actions, including the 'power of things' and the non-human [8], beyond the action of states, boundaries and norms by which we are accustomed to framing phenomena by taxonomies, and failing to grasp their continuous, complex and systemic nature.

2 Decentralised

New infrastructures enabled by emerging digital technologies are allowing new ontologies to be imagined and prototyped, reformulating concepts such as citizenship no longer tied to territorial jurisdictions, for example, but to participation in distributed virtual communities [9]. They are rapidly connecting millions of people into a deterritorialised social network.

According to Orgad and Bauböck [10], global interconnectedness, the recent onset of the concept of a-territorial identity—citizen of the world—and shared responsibility for problems no longer having a local but planetary scale, is giving substance precisely to the concept of global citizenship.

[1] Collier and Aihwa, *Global Assemblages, 12.*

As explained by Craggs and Rashid [11] merging technologies such as blockchain enable the recording of information and transactions without the need for trust between parties, through automated rules in code modified with democratic logic, operated by algorithms.

Those who have seen in the onset of Bitcoin the possibility of changing the financial system for the better (for the purposes of this discourse, net of negative aspects), likewise see these technologies as the promise of a tool to create new institutions [12], not necessarily substitutes for the old centralised, but as decentralised alternatives to existing ones. From infrastructure for managing the commons, to infrastructure that defines new ways of belonging and participating [13].

These approaches certainly lead to a reflection on the break with earlier political theory but shift the focus to its legacies such as separation of powers and theories of sovereignty, leaving us to imagine a strategy that can combine centralisation to protect decentralised activities [14].

According to Sassen [15], the decentralisation of citizenship means that we are witnessing the emergence of a denationalised, transnational, and post-national citizenship. This is an evolution towards an idea of citizenship that wants to remove itself from belonging to a state that defines itself through fear, war, and antagonism [16].

Although the concerns of those who record how numerous decentralised projects have often led to centralised results are well-founded [17], the application of these new infrastructure-institutions to concepts such as temporary citizenship is very promising, precisely because on the one hand the meaning of 'temporary' implies the need for an agile and rapid recodification of the ways in which status is enabled and recognised, and on the other hand enables communities based on belonging to extraterritorial, distributed and decentralised sets of values and affinities.

References

1. Stephens, A.C.: Citizenship without community: time, design and the city. Citizsh. Stud. **14**(1), 31–46 (2010). https://doi.org/10.1080/13621020903466282
2. Deleuze, G., Guattari, F.: A thousand plateaus: Capitalism and schizophrenia. University of Minnesota Press (1987)
3. Murray Li, T.: Practices of assemblage and community forest management. Econ. Soc. **36**(2), 263–293 (2007). https://doi.org/10.1080/03085140701254308
4. Deleuze, G., Parnet, C.: Dialogues. Columbia University Press (1987)
5. Müller, M., Schurr, C.: Assemblage thinking and actor-network theory: conjunctions, disjunctions, cross-fertilisations. Trans. Inst. Br. Geogr. **41**(3), 217–229 (2016). https://doi.org/10.1111/tran.12117
6. Collier, S.J., Ong, A.: Global assemblages anthropological problems. In: Ong, A., Collier, S.J. (eds.) Global Assemblages, pp. 3–21. Blackwell Publishing Ltd (2007). https://doi.org/10.1002/9780470696569.ch1
7. Delanda, M.: New Philosophy of Society: Assemblage Theory and Social Complexity. Bloomsbury Publishing (2006)
8. Bennett, J.: Vibrant Matter: A Political Ecology of Things. Duke University Press (2010)
9. Rheingold, H.: The Virtual Community: Homesteading on the Electronic Frontier (Revised Edition). MIT Press (2000)
10. Orgad, L., Bauböck, R.: Cloud Communities: The Dawn of Global Citizenship? [Working Paper]. https://cadmus.eui.eu//handle/1814/55464 (2018)
11. Craggs, B., Rashid, A.: Trust beyond computation alone: human aspects of trust in blockchain technologies. 2019 In: IEEE/ACM 41st International Conference on Software Engineering: Software Engineering in Society (ICSE-SEIS), 21–30 (2019). https://doi.org/10.1109/ICSE-SEIS.2019.00011
12. De Filippi, P.: Citizenship in the era of blockchain-based virtual nations. In: Bauböck, R. (ed.) Debating Transformations of National Citizenship, pp. 267–277. Springer International Publishing (2018). https://doi.org/10.1007/978-3-319-92719-0_48
13. Rozas, D., Tenorio-Fornés, A., Díaz-Molina, S., Hassan, S.: When ostrom meets blockchain: exploring the potentials of blockchain for commons governance. SAGE Open **11**(1), 2158244021100 2530 (2021). https://doi.org/10.1177/2158244021 1002526
14. Reijers, W., Wuisman, I., Mannan, M., De Filippi, P., Wray, C., Rae-Looi, V., Cubillos Vélez, A., Orgad, L.: Now the code runs itself: on-chain and off-chain governance of blockchain technologies. Topoi **40**(4), 821–831 (2021). https://doi.org/10.1007/s11245-018-9626-5

15. Sassen, S.: Towards post-national and denationalized citizenship. In: Handbook of Citizenship Studies, pp. 277–292. SAGE Publications Ltd (2002). https://doi.org/10.4135/9781848608276
16. Holland, E.W.: Nomad citizenship and global democracy. In: Deleuze and the Social. Edinburgh University Press (2006). https://doi.org/10.3366/edinburgh/9780748620920.003.0010
17. Schneider, N.: Decentralization: an incomplete ambition. J. Cult. Econ. **12**(4), 265–285 (2019). https://doi.org/10.1080/17530350.2019.1589553

Andrea Cattabriga is a PhD candidate in Architecture and Design Cultures at Alma Mater Studiorum University of Bologna (XXXVI Cycle). His Ph.D. research, entitled "(Systemic) Relational Insights. A community-in-the-loop AI approach to generate research insight at systemic scale", investigates an innovative design framework applying hybrid human-artificial intelligence strategies with community-in-the-loop approach, capable to deliver actionable research insights in complex, multi-stakeholder and ecosystemic contexts.

Transition/Deviance

Valentina De Matteo

Abstract

Social structures are living a significant phase shift from complicatedness to complexity and they must exercise their plasticity to cope with the continuous movement from one place to another and the rapid changes of context that require them to adopt always different models of knowledge, interaction and behavior. This attribute of "permanent change" has a deep influence on the meaning of "society" and its founding pillars including that of "citizenship", whatever permanent or temporary it is. The following essay aims to investigate temporary citizenships as a "transition", meaning a series of interconnected changes that take place in different environments but can influence each other and generate continuous innovation. How can the temporary citizen contribute to produce innovation and modify knowledge processes, rituals and approaches within a given territory, a context, a community made of diversity, divergence and difference as potentials traits of deviance?

Keywords

Complexity · Transition design · Open system · Practice · Transition cities · Material citizenship · Diversity · Interdisciplinarity · Democracy of knowledge

1 Transition

In the Treccani dictionary, the noun 'transizione' has at least nine definitions [1]. Music, physics, chemistry, lithology, metallurgy, histology and molecular biology are the different disciplines where the word "transition" has a specific meaning. In a broader sense, transition means a "change from a way of being or living to another, from one condition or situation to a new and different one" [1].

The word therefore has an intrinsic reference to the history of human and social evolution where "different forms of life and language meet with reciprocal influences". The theme of reciprocal influences is a constant feature of transitions: one example is, in a strictly physical sense, the "phase transition" where the transition from one state to another creates an exchange of energy with other systems, to disturb the reference system. From order to disorder to create a new order: this is essentially the effect of all transitions that have become relevant in the studies of our contemporary world's phenomena. A "long" or "prolonged" present is the time

V. De Matteo (✉)
Department of Architecture, Advanced Design Unit, University of Bologna, Bologna, Italy
e-mail: valentina.dematteo2@unibo.it

A. Borsari et al. (eds.), *TEMPORARY: Citizenship, Architecture and City*, The City Project 4,
https://doi.org/10.1007/978-3-031-36667-3_14

frame of "transition design" trying to translate the sense of the "long now" [2] whose goal is "to draw on the knowledge and wisdom of the past to conceive solutions in the present with future generations in mind" [3].

If the "prolonged present" is the time span where transition design reflects and acts, we could consider "cosmopolitan localism" [4, 5] as its spatial frame where the focus on *hic et nunc* never loses the global dimension thanks to a continuous exchange of information and technology. Already in 2015, on the Pisa Internet Festival, Manzini subtitled his speech on cosmopolitan localism as a "co-production of places and communities in a fluid world" [6]. His theory was grounded on the role of connectivity that has dissolved traditional social forms by giving impetus to new ways of co-living and collaborating in new communities to co-design new places. In its evolutionary and adaptive pathway, the city continually becomes a "new place": thanks to the continuous interaction of its actors, we witness the realisation of practices and processes of continuous social learning that demonstrate a deep link between the social and cultural innovation of a territory. This link is also underlined in the general theory of systems whose definition is "a complex of parts which, with certain connotations, establish relationships among themselves, so that the behaviour of each of them is influenced by the relations they establish and are involved in and vice versa" [7]. According to von Bertalanffy, systems are open to, and interact with their environment, and can even acquire qualitatively new properties through emergence, thus they are continuously evolving.

Cities, too, can be considered "open systems" where transitions are constantly informed by new and regenerated knowledge of the natural and social context. An example of concrete application of the concept of transition to cities is the European Union's "Transition Cities" project, of which the city of Bologna is a partner, helping cities move towards a vision of a participatory, sustainable and inclusive future. Participation and inclusiveness within cities also provide for the involvement of "temporary" actors who interact in a network of spaces and relationships and therefore escape the traditional category of "formal citizenship" to converge in a broader concept of "material citizenship". *Practicing* citizenship, therefore, even temporarily, seems to be the way to increase the productive thinking and ability of the city, driven by participation and collaboration. These can be described as "dimensions of change" and are common patterns to the various forms of innovation that we can observe today in the economy and society. The entry "Future of Innovation" in the "Encyclopaedia of Creativity, Invention, Innovation and Entrepreneurship" [8] indicates participation and collaboration as essential dimensions of innovation because they enable the involvement of a wider spectrum of actors and instances. The trend towards a further opening of the innovation process will continue and become even stronger in the coming years: flexible working models and outsourcing, and an ever-increasing number of freelancers who will inhabit changing digital and physical spaces as temporary citizens, are already enabling the emergence of new strategies for innovation. Coordination mechanisms of self-organised user communities, web-based co-design platforms or city-level innovation initiatives involving public and private actors are on the rise. They are progressively becoming complementary to (or replacing) traditional market mechanisms in a scenario of increasingly distributed innovation where the city and its citizens, however stable or temporary they are, will play a key role as a hub of innovation and continuous knowledge production.

2 Deviance

It was 1998 when Fincher and Jacobs defined the "city of difference" as the place where it is possible to perceive the "presence of otherness". In her "*Life and death of large cities. An essay on American metropolises*" published in 1961 [9], Jane Jacobs opened a substantial debate on twentieth-century urbanism and its role in the evolution—or involution—of contemporary society. Density and complexity are the pillars of the urban community which enable four essential rules: the presence of the greatest number of basic functions within the block (housing, commercial activities, businesses, services), the small size of the blocks, buildings of different ages and condition and a good population density (approximately at least 100 units per half hectare). The primary objective of this design is to encourage as much as possible encounters between people who are different in terms of culture, geography, ethnic group and age. These rules are different from the choices made by rationalism: buildings and blocks of maximum size, homogeneous areas for building types and sufficiently low density to prevent hygienic and social problems as issues compromising the established order. We could define these issues as "deviance", the definition of which has been widely debated in both social and sociological terms.

The main difficulties in coming up with a univocal definition of "deviance" are largely because it "is not an objectively given quality of the act, but a subjective (or political) attribution: it is always the product of an interactive process" [10]. In the attempt to find some recurrences in the various definitions of the phenomenon, it is useful to proceed by juxtaposition [11] and by negation [12]. From a sociological point of view, Matza identifies three recurring juxtapositions: correction/compassion (studying the phenomenon to eradicate it *vs* empathically understanding the phenomenon); pathology/diversity (deviance as an intolerable and undesirable variation of normal *vs* deviance as a tolerable phenomenon which is considered negatively only for extrinsic reasons); simplicity/complexity (deviance as an obvious fact breaking the established rules *vs* a complex phenomenon to be defined given its relationships with "normality"). In terms of negation, on the other hand, Howard Becker rejects three conceptions of deviance: the one that reduces it to a pathological state, the one that describes it in statistical terms, and the one that identifies it as transgressive conduct. The quality of the deviance for the author therefore lies in the social reaction and is to be attributed to the eye of the observer. This analysis shows the evolutionary path of the studies on the concept of deviance which has progressively become more scientific and less moralistic starting from the School of Chicago, also known as the School of Urban Social Ecology, where the relationship between deviance and city has been investigated. In Cohen's reconstruction on the sociology of deviance, for the Chicago School the city was seen as a privileged laboratory for studying the phenomenon for ecological reasons (urban growth creates favourable environments for the birth of symbiotic relationships between various forms of deviance), cultural transmission (deviant norms are transmitted and learned from one generation to another), and social disorganisation (deviance is the effect of the weakness of traditional forms of social control).

The debate on the equivalence of meaning between "deviance" and "diversity" is still open but their close relationship has been explicitly investigated by Tamar Pitch [13] in her treatise on deviance. Pitch writes: "if until a few years ago the trend was to interpret this whole complex field by privileging the criminological and pathological aspect, now … we tend to extend the perspective of 'diversity'". For Pitch, it seems that "at the end of its historical journey, the concept therefore tends to approach the notion of 'diversity'. The possibility of accepting new norms leads to the production of new and "hybridised" knowledge, which is nurtured by the concept of "interdisciplinarity". With a pinch of provocation, we could compare the theme of interdisciplinarity to a sort of "deviant behaviour" towards the hyper specialisation of knowledge which has characterised social and organisational contexts for a long time. As Bunders [14] argues, when talking about "democracy of knowledge", interdisciplinarity is not applicable to disciplines only but is a scalable concept that connects the interfaces between science, governance and society to the advantage of solving complex problems. Interdisciplinarity, then, must be considered to all intents and purposes as a mode of innovation because it creates a "feasible interconnection between different levels of cooperation" aimed at improving not only the present but also future development scenarios.

References

1. https://www.treccani.it/vocabolario/transizione/
2. Brand, S.: The Clock of the Long Now: Time and Responsibility. Basic, New York (1999)
3. Irwin, T.: Transition design: a proposal for a new area of design practice, study, and research. Des. Cult. **7**, 229–246 (2015). https://doi.org/10.1080/17547075.2015.1051829
4. Manzini, E.: SLOC: The emerging scenario of small, open, local, connected. In: Harding, S. (ed.) Grow Small, Think Beautiful Ideas for a Sustainable World from Schumacher College, pp. 216–231. Edinburgh, Floris Books (2011)
5. Sachs, W.: Planet Dialectics. Explorations in Environment and Development. Halifax, NS, Fernwood Publications (1999)
6. http://2015.internetfestival.it/wp-content/uploads/2015/10/Localismo-Cosmoplita-EZIO-MANZINI.pdf
7. von Bertalanffy, L.: General System Theory. Foundations, Development, Applications. George Braziller, New York (1968)
8. Leitner, K.-H.: Future of innovation. In: Carayannis, E.G. (ed.) Encyclopaedia of Creativity, Invention, Innovation and Entrepreneurship. Springer, Cham. https://doi.org/10.1007/978-3-319-15347-6_200026 (2020)
9. Jacobs, J., Fincher, R.: Cities of Difference. Guilford Press, New York, NY (1998)
10. Cohen, S.: The critical discourse on social control. Int. J. Sociol. Law. **XVII**, 347–357 (1989).
11. Matza, D.: Becoming deviant. Englewood Cliffs, N. J. (1969)
12. Becker, H.S.: Outsiders: studies in the sociology of deviance, New York (1963)
13. Pitch, T.: Teoría de la Desviación Social, Nueva Imagem (1979)
14. Bunders, J.F., Broerse, J.E., Keil, F., Pohl, C., Scholz, R.W., Zweekhorst, M.B.: How can transdisciplinary research contribute to knowledge democracy? In: In't Veld, R. (ed.) Knowledge Democracy. Springer, Berlin, Heidelberg. https://doi.org/10.1007/978-3-642-11381-9_11 (2010)

Valentina De Matteo is Ph.D. candidate in Advanced Design at Alma Mater Studiorum University of Bologna (XXXVI Cycle). Her Ph.D. research is focused on investigating the role and contribution of design in organizational transformation and capacity building, at the insersection of theories and practices of design, management and organizational sciences.

Hacktivism/Sexual Tourism

Francesco Di Maio

Abstract

The first part analises the problem of temporary citizenship through the case of !Mediengruppe Bitnik's *Random Darknet Shopper*, a piece of hactivism art that has been initially arrested by police, then released. The semiotic concept of discursive regimes is here used to inspect the conflicts between the legal, the artist and the technical istances that are involved. This phenomenon allows to emphasise the discord that emerges between a performative and a right-based citizenship. It also shows how the concept of citizenship, if understood in this paradigm, assumes an extension that can include otherwise unforeseen instances, such as masses of data and goods. The second part develops the concept of sex tourism, attempting to show that this species of tourism is indistinguishable from its genus. Both, in fact, involve and are fuelled by urban zooning phenomena. In the case of sex tourism, it is fed by its illegal nature. After briefly overviewing the global politics concerning sex working legislation, the conclusion focuses on the two processes of projecting a metaphysical truth onto the phenomenon and of doubling down on zoning in the psychological introjection by workers.

Keywords

Hacktivism · Computer art · Discursive regimes · Sexual tourism · Sex workers' rights · Zoning

1 Hacktivism

H., a crasis of 'hacking' and 'activism', is a form of militancy that uses hacking to foster political and social change.

In relation to the topic of temporary citizenship, this phenomenon allows us (*i*) to emphasise the conflict that emerges between a performative and a right-based citizenship; (*ii*) to show how the concept of citizenship, if understood in this paradigm, assumes an extension that can include otherwise unforeseen instances, such as masses of data and goods; the particular case, matrix artist, brings forward how (*iii*) art, besides proposing its own works as "citizens", allows to observe processes analogous to itself, in a closed and controlled form; finally, (*iv*) when hacktivism moves to the limit of the legal, a study of casuistry allows us to demonstrate how citizenship is never given *de jure* a priori, but always and only a posteriori, following the occurrence of a material event (reading: performance) within a discursive

F. Di Maio (✉)
Department of Architecture, University of Bologna, Bologna, Italy
e-mail: francesco.dimaio6@unibo.it

A. Borsari et al. (eds.), *TEMPORARY: Citizenship, Architecture and City*, The City Project 4,
https://doi.org/10.1007/978-3-031-36667-3_15

stratification. For these reasons, it is necessary to return to a specific case study, already analysed elsewhere [1].

Random Darknet Shopper (abbr. RDS, 2014–16) is a computer art project by !Mediengruppe Bitnik, consisting of software that buys goods worth up to $100 in bitcoin, randomly chosen on the darknet. During the exhibitions that hosted this work, the programme made one purchase per week and the packages were shipped to the exhibition site and placed in special showcases, which were progressively filled until the end of the exhibition. Significantly, in the first edition at the Kunst Halle in St. Gallen, Switzerland (14/10/14–15/01/15), RDS purchased from the Agora shop, among other things, 120 mg of ecstasy flown in from Germany and a fake passport. As the artists noted with surprise on their website: “The parcel was sent from Germany and crossed the border and customs to Switzerland without any problems” [2]. The problems started on 12 January 2015, after the exhibition had just ended, when Swiss police confiscated the bot and all the merchandise on display. Only on 15 April, after testing that the pills contained MDMA (90 mg, not the promised 120), did the police destroy them—so they stated—returning everything else.

Although there was a drug purchase, this was not intentional. Although there was drug possession, this was for artistic purposes, therefore RDS created a legal void where, by generating an illicit event, it was possible to attribute an author as the responsible party. Only the artistic framework could make this gesture unpunishable.

The news immediately made the headlines and it was joked that the computer had a human personality and was a legal subject of rights and duties. The Guardian and CBSN babbled that police had “arrested—er, confiscated” a computer [3]. For the Daily Mail they had “ARRESTED” a “cyber criminal!” [4], while the Independent felt it necessary to specify how “the consumer behind these purchases is not actually a human, though—it is an internet ‘bot’” [5]. RDS thus seems to be a full subject, who desires, buys, and commits crimes like all of us: this is the new thing on which it blatantly opens up [6].

The ethical/moral question was immediately raised [7]. Carolina Alves de Lima Salge and Nicholas Berente proposed establishing the moral value of the bot through an algorithm, i.e., an algorithm of a higher order than the one of the software, thus opening up the risk of regression to infinity [8]. The authors ask: although a state law was violated (the bot bought drugs banned on the regular market and a false passport), can this behaviour be considered “unethical”? Their meta-algorithm consists of three questions: (*i*) Did the social bot’s action break the law? (*ii*) Is its action unjustifiable? and, finally, (*iii*) Is there any deception involved? A bot, but in general any moral subject, does behave unethically if its action answers affirmatively in progression to the three questions. The authors’ conclusion and aim—a full-fledged categorical imperative—is that “social bots should always act truthfully” [9].

Since the purpose was artistic and the ecstasy “in this presentation was safe”, as indicated by the police, the authors of the article can consider the bot’s behaviour as “not unethical” since it is justifiable by the widespread morality of the community. Here, though, the impasse: for the two scholars, who refer to Aristoteles’ *Et. Nic.* III, 1111a, an ethical subject-bot emerges as such, if and only if its action is both known and intended. RDS, however, by making random choices, acts as alea, to which it is fictional to attribute any awareness of will. Thus, RDS moves alongside the meta-algorithm, in the space of a legal void.

All that remains is to blame the developers [10], but again, this is a special solution which is not always valid. The substance of the subject is attributed to the signatory, as the first fixed reference in the chain of statements that has been created. It is precisely by signing that a fixed and constant instance is established, even though the signature itself indicates the authors’ extraneousness to the facts and their non-presence. In that case there would be no fault, at most, fraud. A conflict thus arises between two discursive regimes, on the one hand, that of the law whose purpose is to interrupt referrals of enunciation [11], such as a signatory that has emerged retroactively to their signature [12]; on the other, the discursive regime of technique, where each actor's production amplifies the chain of references with other actorly projections [13]. Finally, copyright allows the Swiss duo Carmen Weisskopf and Domagoj Smoljo to claim the genesis of the work for themselves, taking responsibility as developers [14]. The artistic framework [15] ultimately establishes a third discourse that makes it possible to neutralise any kind of guilt on the part of RDS or its authors.

In conclusion, RDS presents itself as an emblematic "example of hacktivist guerrilla communication", the sole *raison d'être* of which is "reflection on the shadowy parts of the Internet without calling for any specific legal change or articulating any specific political claim" [16]. RDS does not want to bring about change in any current law, but to raise a reflection on the limits of the legislature itself.

2 Sexual Tourism

> [...] Tourism, a sad invention. There is nothing outside your own cave. Standing still.
> Tourism, sexual impulse. When one knows oneself well, on this side, it is useless to move; or to move for work, working [17].

S. t. refers to the group of tourists present, in a specific period of time, in a country, in an area, in a city, for the purely material purpose of entertaining erotic relations [18]. However, it is difficult to distinguish this specific type of t. from its analogues.

The nature of international mass tourism, supposedly globalising and homogenising, finds four related words in the four "S's" of "Sun, Sand, Surf and Sex" [17, 19], which arrive at a point of indistinction insofar as sex can become an "accessory" of cultural or beach tourism, or vice versa [18, 20].

The stereotypical sex tourist figure is a white, Western man who travels to Asia to procure the sexual services of a young girl, a stereotype that doubles as western women who can afford to pay for sex [19, 21]. Destinations are specific: "to Thailand or South East Asia more generally, hundreds of thousands of Western women to the Caribbean or to Gambia, and (uncounted) multitudes of gay men to the Maghreb", as well as in Europe, where cities such as Budapest and Prague specialise in this type of tourism. The stereotype is understandable given its resonance, which has its origins in other events: an example is in Thailand where the sex industry for foreigners was spawned by the Vietnam War, during which Bangkok, as a behind-the-lines retreat, "began to specialise in the entertainment of American soldiers and marines on authorised leave". According to some estimates, there are "around 2.8 million sex workers, 800,000 of whom are minors" [22].

The relationship between destination and sex tourism reacts with that between sex working and legality. The latter can be summarized in the following categories:

- Prohibitionism: the purchase of sexual services is criminalised, following the Lombrosian concept that those who prostitute themselves are dangerous (Asia, Africa, and the Middle East, USA, outside of Nevada);
- Regulationism: sex working is a necessary evil; the state intervenes in a massive and violent way, but does not recognise it as work (Turkey, where those who work are filed and registered; you can work only if you are not married);
- Abolitionism: prostitutes are victims, prostitution is legal but not recognised as work and there are collateral crimes related to the practice of the profession (Argentina; England; Spain; Italy);
- Neo-regulationism or legalisation: prostitution is regulated, some rights are protected, at least in theory, and a legal profile to the activity is recognised, although it is distinguished from other jobs (Germany; Switzerland; Holland);
- Neo-abolitionism: sex workers are women victims of patriarchy and those who buy are executioners to be punished. Purchasing is forbidden, in the belief that this solution will lead to the elimination of sex work; it is a violent model with serious consequences on workers (Sweden; Norway; Canada; Iceland; France);

– Decriminalisation: the job has only been recognised since 2003 and can be exercised both inside and outside the home, both in authorised centres and in apartments, although this model must apply only to recognised citizens and not migrants (New Zealand; Australia (New South Wales and Northern Territory) [23].

Considering the intrinsic relationship with the illegal, this type of consumption, developing in extension, takes on the characteristics of tourism in general, confirming the indistinguishability of genus and species: indeed, on the urban level, tourism is grafted as a *dispositif* that separates residents from visitors, disproving the anthropological category of the relationship "guests and invited". The two groups brush against each other rather than touch, phenomenon described by the Chicago school as a "processes of [urban] segregation" that in the big city generates a "mosaic of little worlds which touch but do not interpenetrate", "contiguous, but otherwise widely separated" [24]. Urban space physically manifests this division between "natives" and "tourists" through processes of zoning, i.e. dividing the city into zones, each assigned to a different function, the first of which was New York City in 1916 [25]. By creating an autonomous economy that does not reinvest in the territory from which it extracts, it leads to disillusionment with any form of sustainable tourism [26] whether sexual or not.

Zoning, by bringing tourism to inner but segregated areas, has mantled the places of consumption with a value of transcendence to the eye of the customer, to the extent that we have relied on senses other than sight. As poststructuralism has highlighted, it is about the gap between regimes of visibility, between what is visible and what can be said about it. Speeches pertain only to certain saliences and exclude the rest.

As the exterior is constituted as a fold of the interior, a specific metaphysics has traditionally been advanced by a male gaze towards the prostitute, as an image of truth [27], or even a theology, as in the novels from Georges Bataille [28]. In the '50 s, Tommaso Landolfi, admitted how "we didn't feel that we really knew a city until, like dogs in the street, we had put our nose in one of 'those little places'" [29]. As Laurent de Sutter summarises, at the basis of which is the historiographic reconstruction of Amélie Maugére [30]:

> *The public is the truth of the private*: the inside is the truth of the outside – but an inverted, invaginated truth, turned inside out in the manner of a glove finger, that figure of topological continuity. The history of urban policy on prostitution has always favoured the discrete: the establishment of a solution of continuity between the places where whores are admitted, and those where they are not welcome. The whorehouse, as a historical invention, is the most famous of these solutions of continuity, although it is not the only one: infamous marks or reverse districts have also played the role of suture [31].

The urban topology doubles in the so-called "fronteras *encarnadas*" ("*incarnate* frontiers"), that is, in strategies on the part of self-protection to separate work and private spheres: setting a limit to certain practices, defining the duration of the service, dissociating mentally and physically from the sexual act, preserving an emotional detachment by not experiencing pleasure [32].

References

1. Di Maio, F.: Alea e anomia. Analisi gius-filosofica di un caso di computer art. Philosophy Kitchen (2021), https://philosophykitchen.com/2021/05/alea-e-anomia-analisi-gius-filosofica-di-un-caso-di-computer-art/, last accessed 2022/27/10; Id.: Ibridi ed enunciazioni. Attorno a Bruno Latour. In Ghelfi, A. (ed.) Connessioni precarie. Per una ecologia politica della rigenerazione: leggendo Haraway, Stengers e Latour, pp. 72–83. ombre corte, Verona (2022).
2. !Mediengruppe Bitnik: random darknet shopper (2014–2016), https://www.bitnik.org/r/, Last accessed 27 Oct 2022
3. Kasperkevic, J.: Swiss Police Release Robot That Bought Ecstasy Online. The Guardian (2015) https://www.theguardian.com/world/2015/apr/22/swiss-police-release-robot-random-darknet-shopper-ecstasy-deep-web, Last accessed 03 May 2022; Kharpal, A.: Robot with $100 bitcoin buys drugs, gets arrested. CBSN (2015), https://www.cnbc.com/2015/04/21/robot-with-100-bitcoin-buys-drugs-gets-arrested.html. Last accessed 03 May 2022

4. Grey, R.: Now that's a cyber criminal! Robot is arrested by police for buying ecstasy on the dark net. Daily Mail (2015), https://www.dailymail.co.uk/sciencetech/article-3047317/Now-s-cyber-criminal-Robot-ARRESTED-police-buying-ecstasy-dark-net.html, Last accessed 03 May 2022
5. Grant, K.: Random Darknet Shopper. Exhibition featuring automated dark web purchases opens in London. The Independent (2014), https://www.independent.co.uk/life-style/gadgets-and-tech/news/random-darknet-shopper-exhibition-featuring-automated-dark-web-purchases-opens-in-london-a6770316.html. Last accessed 03 May 2022
6. Noto La Diega, G., Walden, I. (2016). Contracting for the "Internet of Things". Looking into the Nest, p. 4 note 15, Queen Mary School of Law Legal Studies Research Paper 219 (2016)
7. Power, M.: What happens when a software bot goes on a darknet shopping spree?. The Guardian (2014), https://www.theguardian.com/technology/2014/dec/05/software-bot-darknet-shopping-spree-random-shopper. Last accessed 03 May 2022
8. Shaw, J.: Evil. The Science Behind Humanity's Dark Side. Abrams, New York (2018)
9. Alves de Lima Salge, C., Berente, N.: Is That Social Bot Behaving Unethically? Commun. ACM, **60**(9), 29–31 (2017)
10. *Ivi*, p. 31
11. Latour, B.: Petite philosophie de l'énonciation, http://www.bruno-latour.fr/sites/default/files/75-FABBRI-FR.pdf. Last accessed 03 May 2022
12. Derrida, J.: Signature Event Context (1971). In: Bass, A. (ed.). Id.: Margins of Philosophy (1972), pp. 307–30. University of Chicago Press, Chicago (1982)
13. Latour, B.: *op. cit*
14. Andrini, L.: Redesigning Indonesia copyright act to accommodate autonomous intelligent system. *Status Quo* and Room for Improvement, Asian J. Law Econ., **9**(3), 1–9 (2018); Lagioia, F.: Responsabilità Penale e Automazione nell'*E-Health*, pp. 126 and 129. Tesi di dottorato, Alma Mater Studiorum—Università di Bologna, Bologna, Italia (2016)
15. Danto, A.: The Artworld. Journal of Philosophy **61** (19), 571–584 (1964)
16. Delmas, C.: Is Hacktivism the New Civil Disobedience? Raisons politiques, **69**(1), 63–81
17. Flaiano, E.: Secondo viaggio. Agosto 1950. In: Id.: Diario degli errori (1976), §24, p. 18. Adelphi, Milan (2002)
18. 'turismo', https://www.treccani.it/vocabolario/turismo/. Last accessed 03 Apr 2022
19. Ryan, Ch., Michael, C. H.: Sex tourism. Marginal People and Liminalities, p. ix. Routledge, London (2001)
20. D'Eramo, M.: The World in a Selfie. An Inquiry into the Tourist Age (2017), English translation Bowett-Jones, B, Broder, D., p. 57. Verso, London (2021)
21. Murray, A.: Tourism. In: Ditmore, M.H. (ed.) Encyclopedia of Prostitution and Sex Work, vols. 1–2, p. 488. Westport, London, Greenwood (2006)
22. D'Eramo, M.: *op. cit.*, p. 58
23. Weitzer, R.: Criminalization, Decriminalization, and Legalization, pp. 375–402. In Id. (ed.) Sex for Sale. Prostitution, Pornography, and Erotic Dancing, Routledge, New York (2023^3); Zollino, G.: Sex work is work, pp. 18–9. eris, Torino (2021)
24. Park, R. E.: The City. Suggestions for Investigation of Human Behavior in the Urban Environment (1928). In Id., Burgess, E. W., McKenzie, R. D.: The City, pp. 40–1. Chicago UP, Chicago (1967)
25. Moskowitz, M.: Zoning the Industrial City. Planners, Commissioners, and Boosters in the 1920s. Business and Economic History 27(2), p. 307 (1998)
26. Gainsforth, S.: Oltre il turismo. Esiste un turismo sostenibile? eris, Turin (2020)
27. De Sutter, L.: Métaphysique de la Putain. Scheer, Paris (2014)
28. Bataille, G.: Madame Edwarda (1937), pp. 223–36. In Id.: The Bataille Reader, eds. Botting, F., Wilson, S. Blackwell, Oxford (1997)
29. Landolfi, T.: Se Non la Realtà (1960), p. 115. Adelphi, Milan (2003)
30. Maugére, A.: Les Politiques de la prostitution. Du Moyen Age au XXIème siècle, Dalloz, Paris (2009)
31. De Sutter, L.: *op. cit.*, § 27
32. Morcillo, S.: De cómo vender sexo y no morir en el intento. Fronteras *encarnadas* y tácticas de quienes trabajan en el mercado sexual. Revista Latinoamericana de Estudios sobre Cuerpos, Emociones y Sociedad **3**(7), 17–28 (2011)

Francesco Di Maio is a Ph.D. candidate in Architecture and Design Cultures at Alma Mater Studiorum University of Bologna (XXXVI Cycle), in cotutorship with the *école doctorale* Droit, Études Politiques, Philosophie (DEPP) of École des Hautes Études en Sciences Sociales (EHESS) in Paris. His research project, entitled "Space, art and nature in F.W.J Schelling", is dedicated to the problem of spatialization and architecture in the aesthetics and in the *Naturphilosophie* of the German philosopher.

Life Cycle Thinking/Life Cycle Phase

Lorna Dragonetti

Abstract

The economic model that has characterised the last 150 years is the so-called "linear model" and the extraction of new material and its decommissioning has led to the current climate crisis. Life Cycle Thinking (LCT) is an approach that looks for strategies to reduce the impact and the consumption of resources by goods and services throughout their entire cycle. LCT is needed to operate even where environmental impacts are forgotten, such as transitory residence situations like tourism, reception buildings for migrants, student/worker accommodation, and social housing. These buildings are rarely renovated and they are often at the end of their natural life, but are used again as a residence, giving rise to a new cycle with its own impact. Each type of approach based on the concept of life cycle provides for interventions, with precise purposes, differentiated for each single phase of an object. In this cycle, the phases alternate one after another, in a continuous transformation of matter and energy that leads to the creation and use of goods and services, and each phase brings with it a profound change in the conditions set in place in the previous one. Similarly, the life of each individual can also be divided into stages, with the most varied characteristics and facets, which mark the passage of years. For sustainable development, it is necessary to be aware that every phase of life of a good or person has an impact and must be studied and adjusted to minimise it.

Keywords

Environmental impact · Life cycle · Circularity

1 Life Cycle Thinking

The economic model that has characterised the last 150 years is the so-called "linear model", based on the use of all new raw materials and the production of waste at a product's end of life. The continuous extraction of new material and its rapid decommissioning has created enormous environmental damage, leading to the current climate crisis that we are all experiencing. Years of research to devise a more sustainable development model have merged into the concept of the "circular economy", a model of production and consumption that regenerates and minimises waste. The key words are reuse, regenerate, repair, and using materials and products for as long as possible [1].

L. Dragonetti (✉)
Department of Architecture, University of Bologna, Bologna, Italy
e-mail: lorna.dragonetti2@unibo.it

A. Borsari et al. (eds.), *TEMPORARY: Citizenship, Architecture and City*, The City Project 4,
https://doi.org/10.1007/978-3-031-36667-3_16

Every year in the EU, about 2.5 billion tons of waste are produced, an unsustainable quantity that must be drastically decreased. For this reason, in February 2021 the European Parliament voted to adopt a "new action plan for the circular economy" for a transition to a carbon–neutral, sustainable and fully circular economy by 2050 [2]. To achieve these goals, it is necessary to adopt approaches that take into account the entire life cycle of materials and products (life cycle approaches) because only by looking at production from this perspective can we minimise the impact in every aspect, including disposal, or better reuse and recycling.

It is necessary to extend the life cycle of products as much as possible. Until recently, most of the environmental improvement actions put in place have been focused on reducing pollutant emissions from individual sources (e.g. discharges into watercourses, processing waste in the industrial sector, emissions of gases and fine particles from transport and factories). Life Cycle Thinking (LCT), on the other hand, is an approach that looks for strategies to reduce the environmental impact and the consumption of resources by goods and services throughout their entire life cycle, starting from the production of raw materials, passing from production, distribution, use and consumption, up to the reuse or recycling of materials, the recovery of energy and the disposal of any waste that can no longer be used [3].

Understanding the environmental impact of the whole life cycle is a relatively new concept that finds its roots in the 1960s and 1970s with the first studies centred on the scarcity of raw materials (especially oil). Those years saw the first studies on the environmental profile and energy analysis that have now become mandatory and an important decision-making tool both in the public and private domain. Today, LCA (Life Cycle Assessment) for environmental impact analysis and LCC (Life Cycle Costing) for economic impact analysis are used to estimate life cycle impacts. These analyses are used to calculate the environmental and economic costs throughout the whole life of a good or service, in order to certify its sustainability or to support the decision-making process.

In a Europe which is increasingly focused on mitigating the effects of climate change, an approach such as LCT is needed to operate even where environmental impacts are forgotten, such as transitory residence situations like tourism, reception buildings for migrants, student/worker accommodation, and social housing. These buildings are rarely renovated and are often old houses almost at the end of their natural life, but are used again as a residence, giving rise to a new cycle with its own impact [4].

In these cases, the LCT approach aims to estimate the environmental, economic, or social impacts of a type of temporary dwelling, variable over time. A temporary condition of citizenship or living may in some cases have a strong environmental impact, but may also lead to a lengthening of the life cycle of dwellings and structures that in other cases would remain unoccupied. For this reason, it is necessary to intervene and understand how the reuse of these goods can be shaped according to the needs of temporary occupants and can be upgraded to last as long as possible to accommodate new people, thereby reducing emissions from their use.

The conditions of migrant shelters or old houses in historic centres rented to students or tourists are often obsolete, with high energy consumption and a not very comfortable living environment. These structures are not usually renovated because the inhabitants live here just for a period of their life, and then the houses welcome others. The resource consumption of each individual period of occupation is relatively low but if you look at the entire life cycle of the building, the amount of waste and emissions becomes enormous and must be handled by the owners and public administrations that should facilitate processes of regeneration of these goods.

2 Life Cycle Phase

Life is a succession of instants, days and years, and some of these fragments of time are perceived (even afterwards) as a single moment or even "phase of life" because they are united by characteristics, goals and sensations. A phase can be defined as "each of the characteristic and differentiated moments of a continuous development" [5] and this succession of moments unites all and everything around us whether people, animals or things. While the life cycle of living beings is linear, with a beginning (birth) and a coming to an end (death), the life cycle of objects can be thought of as "circular", where instead of an end there is a rebirth, a reuse and a new beginning that constitutes the new "phase 0" [6].

It is precisely this new circular approach that must be adopted for sustainable development and minimising of loss, waste and emissions in every single phase of life of an object or a service (see Life Cycle Thinking), through the so-called "life cycle approaches". Each type of approach based on the concept of life cycle provides for interventions, with precise purposes, differentiated for each single phase of an object: production, distribution, use, disposal, recycling.

In this cycle, the phases alternate one after another, in a continuous transformation of matter and energy that leads to the creation and use of goods and services, and each phase brings with it a profound change in the conditions set in place in the previous one. Similarly, although with a linear trend, the life of each individual can also be divided into stages (birth, childhood, youth, adulthood, aging, death), with the most varied characteristics and facets, which mark the passage of years.

Sustainable design also borrows the terms from this analogy with human life and indeed there are two distinguished approaches, "from cradle to grave", (i.e. linear) and "from cradle to cradle" (i.e. circular), preferring the latter where the waste material is incorporated back into the production cycle and gives life to a new beginning.

As well as in the productive cycles, to transform matter it is required to absorb energy; in order to evolve and face changes, people must invest energy in their emotional, physical and relational experience. During this continuous growth, changes arise from the interaction between the individual and the environment, and this development is made of acquisitions and losses, crisis and continuity. Each life cycle, therefore, is a succession of phases, each well-defined with its own characteristics but part of a historical continuum that includes the individual phases and the moments of transition between them. It is often in these moments of transition, in which the crisis of the status quo takes place, that there is the greatest instantaneous consumption of energy, but these are also moments in which small revolutions take place that will determine the course of the next phase.

For many people, moreover, some of these stages of life may involve more or less prolonged or definitive displacements from one's place of origin. Whether changes of address within the same city, or transfers from one corner of the world to another, people are in any case living in a phase of temporary citizenship, a reality that characterises the lives of many. Almost all migrants, tourists and off-site students have gone through this phase in their lives, maybe more than once, treasuring the experiences made and the places visited.

In some cases, individuals develop a strong and rooted sense of belonging to a place, sometimes overcoming that of their origins. In other cases, the best opportunities offered by the place of temporary residence lead to the choice of making such citizenship permanent, creating new settlements and families and triggering almost a new life cycle, that in the case of the human being cannot be circular, but can often start from (almost) zero. Obviously, one cannot think that there is uniformity at this stage of people's lives, indeed, a tourist's experience of temporary citizenship is completely different from that of a migrant and the feelings and conditions of life cannot be compared. The tourist does not live a "phase of life" in their experience of temporary citizenship, while the migrant or off-site student will live a moment that will mark their lives, and in some cases this place will no longer be temporary, instead becoming definitive.

The experience of temporary citizenship is not limited to people, but also applies to buildings that in a moment of their life can be found to accommodate these temporary inhabitants (for example, reception facilities, communities, family homes, houses for rent to students or off-site workers) and which, as often happens in these cases, lose their identity and are left in unpleasant conditions, not being owned by the inhabitants and not being inhabited by the owners.

For sustainable development, it is necessary to be aware that every phase of life of a good or person has an impact on the environment and must be studied, adjusted and adapted so that this impact is minimised, but also to achieve a social equity that is often lacking in these moments of temporary living.

References

1. Erlandsson, M., Borg, M.: Generic LCA-methodology applicable for buildings, constructions and operation services—today practice and development needs. Build. Environ. **38**(7), 919–938 (2003)
2. European Parliament: European Parliament resolution of 10 February 2021 on the New Circular Economy Action Plan (2020/2077(INI)) (2021)
3. EU Commission: Making Sustainable Consumption and Production a Reality—A Guide for Business and Policy Makers to Life Cycle Thinking and Assessment (2010)
4. Barberis, W.: L'impatto Del Settore Turistico Sui Contesti Urbani-Riflessioni Sulla Programmazione Territoriale. Cittalia-Anci Ricerche (2008)
5. Oxford Languages and Google—Italian|Oxford Languages. Last accessed 15 Sept 2021
6. Bach, V., et al.: Assessing the ability of the cradle to cradle certified products program to reliably determine the environmental performance of products. Sustainability **10**(5), 1562 (2018)
7. EU Commission: Making Sustainable Consumption and Production a Reality—A Guide for Business and Policy Makers to Life Cycle Thinking and Assessment (2010)
8. European Parliament: Waste management in the EU: infographic with facts and figures (2018)

Lorna Dragonetti is a Ph.D. candidate in Architecture and Design Cultures at Alma Mater Studiorum University of Bologna (XXXV Cycle). Her PhD research, entitled "New strategies to activate the market of architectural and energy renovation of existing buildings", aims to develop new strategies to promote deep renovation projects for existing buildings with user-oriented approaches.

Time-Based Packaged Goods/*Valigia* (Suitcase, Luggage)

Clara Giardina

Abstract

The first lemma defines what are those packaged goods conditioned by the "time" variable. A people-planet-based design approach opens up a consideration with respect to those goods (and thus packaged goods) where the functional dimension intersects with a temporal dimension on the 'end of life'. The dimension of temporary citizenship and its repercussions on packaged goods is also explored. In the second lemma, on the other hand, the investigation focuses on a particular type of case, the suitcase, which introduces the "space" variable within the research. In addition to an investigation of the object's role over the centuries, we explore how the suitcase intertwines its history with that of migration, becoming the intimate baggage of migrants to carry a constellation of small objects that relate past, present, and future.

Keywords

Packaged goods · Temporal and spatial dimensions · Sustainability · Temporary citizenship · Luggage · Migration

C. Giardina (✉)
Department of Architecture, University of Bologna, Bologna, Italy
e-mail: clara.giardina@unibo.it

1 Time-Based Packaged Goods

Packaging represents the synthesis of mature consumption-based capitalist behaviour as well as the summary of an articulated and complex set of material and immaterial facts through which goods (the commodity) are brought from production to consumption [1].

Packaging, far beyond the mere function of "wrapping the goods", is a complex artifact with multiple purposes, which have expanded over time. "It is an object of use (instrumental prosthesis), which allows the product to be contained, protected, stored, transported and, at the same time, to facilitate its physical interaction with the user. Furthermore, it is a communication device (communication prosthesis), characterised by functions such as by appellation, persuasive, informative and prescriptive functions, among others" [2].

A design approach based on the needs and well-being of people and the planet (people-planet-based approach) opens a reflection with respect to those goods (and therefore packaging) where the functional dimension intersects with a temporal dimension.

Packaging has a temporary use, limited in time, precisely because of the function for which it was designed, linked to a set of products/goods that must be discarded and used.

However, this time dimension undergoes modifications in some contexts. In this sense it is

A. Borsari et al. (eds.), *TEMPORARY: Citizenship, Architecture and City*, The City Project 4,
https://doi.org/10.1007/978-3-031-36667-3_17

possible to isolate two streams among many possible ones.

On the one hand, design interventions aimed at achieving more "sustainable" packaging are, for all intents and purposes, attempts to extend the life of the packaging (or the material of which it is composed)—in itself temporary and short—as long as possible over time.

This first category includes packaging made of compostable, biodegradable, recycled or waste-derived materials [3], services with durable packaging (reusable for the same or different uses) [4] (Fig. 1), and circular solutions [5, 6].

A second strand, which intertwines the functional (but also the communicative) with the temporal dimension, is represented by that set of products/goods (and related packaging) designed for people experiencing situations of "temporary citizenship' or transitory status in general: tourists, migrants, off-site students, as well as citizens crossing the city on foot or by transport. Many of these products also simultaneously embrace a sustainable approach.

In this cluster we find products designed for staying in hotels, services and packaged foods designed to create a bridge between the place of origin and the place of temporary stay (Fig. 2); products that hook into the services offered by the city (Fig. 3), packages that enhance the experience of eating on the street or at home (a category stressed by the restrictions caused by Covid in recent years), kits and services designed for those temporarily inhabiting a means of transportation (such as the Trenitalia Health & Safety Kit consisting of mask, sanitising gel and water to the traveller to make the trip more comfortable).

This set of product packages that emphasise the temporal dimension highlights an informational, communicative and formal language that addresses both the contemporary citizen (mindful about sustainability practices) and the temporary one (transiting, or moving through space and time), addresses their needs and desires, and develops innovation from them.

2 Valigia (Suitcase, Luggage)

A container equipped with handles or grips for storing clothing and other personal and useful items to take with you on a transit, the "valigia" (Italian word for luggage) is an object generally shaped like a parallelepiped, and a word of uncertain etymology: it could derive from the Arabic ualiha (sack), the Latin vidulus (suitcase, precisely) or the German Felleisen (animal skin) [7].

In the past, the word had broader meanings, indicating other means and modes of transporting objects while travelling. For example, in the days of diligence (horse-drawn carriages used for scheduled transport of travellers and luggage), the bag in which correspondence and parcels were stored was called suitcase or postal case [8].

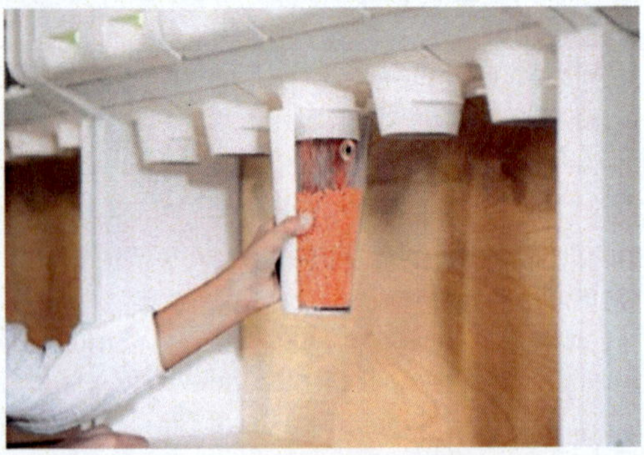

Fig. 1 MIWA is an ecosystem for smart and reusable packaging, based on a clean and effective way to sell goods while minimising packaging waste. *Credits* Minimum Waste|Miwa www.miwa.eu

Fig. 2 Daggiù (literally "from the south"), a surprise food box featuring a selection of southern Italian delicacies. *Credits* Daggiu srl

Fig.3 Phil (Palomar), a water bottle that shows you where to find the city's water fountains/water supply points. *Credits* Palomar www.palomarweb.com

In 1596, the word 'luggage', meaning of awkward and heavy baggage, was added to the Oxford English Dictionary. The word derives from the verb "to lug", meaning to pull, to carry.

The use of suitcases to transport objects has ancient origins. According to some accounts, the first to make use of wheeled luggage were the Knights Templar, to carry armour and other items during the Crusades [9]. From informal luggage made of leather, cardboard or fabric, it became a more durable trunk with squared-off shapes (more suitable for stowing on a ship) thanks in part to the intuition of a young Parisian manufacturer around 1870—Louis Vuitton—who transformed it into "the packaging of fashion" that "safely packs the most fragile objects", as Vuitton's first manifesto read. In contrast, it was in 1974 that Bernard Sadow patented the wheeled luggage, an ingenious intuition that struggled to catch on for many years and is instead representative of the traveller/tourist of the present time [10].

Well before becoming an object related to travel in the contemporary sense, the suitcase intertwines its history with that of migration, and with the individual stories of migrants who for reasons of economy, war, survival, and achievement leave their homes behind and carry a piece of their history in a suitcase.

From the migrant landings at Ellis Island in the United States in the 1950s to more recent migrations—but also in any kind of temporary displacement where the subject goes to inhabit a different space for a more or less defined time—the suitcase (together with its contents, meticulously chosen) becomes a "vector of subjectivity", a "treasure chest of desire", full of the "imagined time, which re-articulates present and future" as Selenia Marabello defines it in *Oggetti Contesi. Le cose nella migrazione* [11].

Baggage or what remains of it, at the end of a migration journey, also has to do with the possibility, in the new place, of "making home", with reference to that "mechanism of organising one's space" that makes a place more one's own, as Pisoni writes in *Il bagaglio intimo. Gli oggetti dei migranti in viaggio verso l'Europa* [12]. In this book, Pisoni uses the ethnoarchaeological lens for his research, going so far as to "emphasise the relational and phenomenological component that characterises both the use of the objects brought for the journey and the practices enacted in the host structures in Europe". Objects (of a religious, sporting or nationalistic nature) certainly have a symbolic meaning, but "the relational system that links the material culture to the context left behind at the time of departure, determining the choice of objects to put in the luggage and influencing the practices of homing and storytelling that, with an eye to the past, attempt to reconfigure the migrants' compass toward the future" also deserves attention.

As in Charles Dickens' *The Anonymous Baggage*, [13] each object is a story, a tale, a mystery unfolding. The intimate baggage of migrants is thus a constellation of small objects (crucifixes, holy cards, Bibles, Korans, bracelets, amulets, make-up, clothing, traditional clothes, family photos and pictures, smartphones) that have been chosen and travelled with them, a material ensemble that constitutes the "personal facade" of the subjects [14] and interrelates past, present and future.

References

1. Celaschi, F.: Verso la fondazione di un Osservatorio Permanente sull'innovazione nel packaging. Internal Report, University of Bologna (2019)
2. Ciravegna, E.: Diseño de packaging. Una aproximación sistémica a un artefacto complejo. RChD: creación y pensamiento (2017)

3. Circular Design Project: Design For X Tool. http://www.circulardesign.it/design-for-x/# Accessed 27 May 2022
4. Ellen Macarthur Foundation Reuse Guide. https://www.ellenmacarthurfoundation.org/assets/downloads/Reuse.pdf
5. Ellen MacArthur Foundation and McKinsey & Company, Towards the Circular Economy Vol. 2: Opportunities for the Consumer Goods Sector (2013)
6. Circular Design Guide Article. https://www.circulardesignguide.com/story/design-has-to-be-driven-by-purpose Accessed 27 May 2022
7. it.wikipedia.org/wiki/Valigia
8. treccani.it/vocabolario/valigia/
9. Barrel, S., Simon, K.: How to Pack for Any Trip. Lonely Planet (2016)
10. Mchugh, J.: A Brief History of the Modern Suitcase (2016)
11. Ascari, P.: Oggetti Contesi. Le cose nella migrazione. Mimesis Edizioni (2020)
12. Pisoni, L.: Il bagaglio intimo. Gli oggetti dei migranti in viaggio verso l'Europa, Meltemi, Milan (2018)
13. Dickens, C.: Il bagaglio anonimo. Faligi Editore (2014)
14. Goffman, E.: The Presentation of Self in Everyday Life, Bantam Doubleday Dell Publishing Group Inc (1959) (Italian transl) (1969)

Clara Giardina is sustainable innovation and packaging expert, Ph.D. candidate in Architecture and Design Cultures at Alma Mater Studiorum University of Bologna. Her research concerns the development of an R&I platform for Advanced Packaging Design, with the aim of providing tools and methods that support circular and sustainable innovation through systems thinking.

Adaptation/*Cueva* and *Tomas De Terreno*

Marco Iannantuono

Abstract

In psychology, the term *adaptation* refers to the individual's *adaptation* to the environment, the relationship that the individual establishes with the environment in which he or she acts and operates, so as to be able to obtain the satisfaction of his or her physical and social needs; on the other hand, in biology, it refers to the modification to which organisms are subjected when environmental conditions change: individuals and organisms interacting—and adapting—to the environment. Interaction and integration, moreover, are fundamental aspects of contributing to a civil society, understood as the totality of associative, economic, cultural and social relations between citizens in complex societies. Starting from the debate on 'TEMPORARY Citizenship, Architecture and City', the cue of the interpretation of citizenship—according to distinct connotations—offers an interesting stimulus for research. The lemma proposal aims to study adaptive forms of architecture, according to an interdisciplinary approach, in particular by addressing a preliminary description of Complex Adaptive Systems (CAS), drawing from disciplines such as biology, so as to contemplate different definitions and interactions between various subjects: organisms, individuals and environments. Finally, some examples of 'adaptive architecture' are described.

Keywords

Adaptation · Cueva · Tomas de terreno

M. Iannantuono (✉)
Department of Architecture, University of Bologna, Bologna, Italy
e-mail: marco.iannantuono2@unibo.it

1 Adaptation

There are many different definitions of *adaptation*. For example, in accordance with the Merriam-Webster Dictionary the term *adaptation* refers to the adjustment to environmental conditions, such as: the "adjustment of a sense organ to the intensity or quality of stimulation" or the "modification of an organism or its parts that makes it more fit for existence under the conditions of its environment: a heritable physical or behavioural trait that serves a specific function and improves an organism's fitness or survival" [1].

Furthermore, in psychology the term *adaptation* refers to the individual's *adaptation* to the environment, the relationship that the individual establishes with the environment in which he or she acts and operates, so as to be able to obtain the satisfaction of his or her physical and social

A. Borsari et al. (eds.), *TEMPORARY: Citizenship, Architecture and City*, The City Project 4,
https://doi.org/10.1007/978-3-031-36667-3_18

needs; on the other hand, in biology, it refers to the modification to which organisms are subjected when environmental conditions change: individuals and organisms interacting—and adapting—to the environment [2].

Interaction and integration, moreover, are fundamental aspects of contributing to a civil society, understood as the totality of associative, economic, cultural, and social relations between citizens in complex societies.

As an interdisciplinary domain, complex systems draw contributions from many different fields, such as the study of self-organisation and critical phenomena from physics, spontaneous order from the social sciences, chaos from mathematics, *adaptation* from biology, and many others. Complex systems are therefore often used as a broad term encompassing a multi-objective research approach to problems from many different disciplines.

Complex Adaptive System are special cases in that they have the capacity to change and learn from experience, examples being cities and any human endeavour based on social aggregation in a cultural system.

In the field of contemporary architecture, some recent research projects explore the *adaptation* primarily as a technological system that reacts to changing environmental and climatic conditions. To develop a new holistic and resilient approach for the future [3].

Recent technological developments in biology, computation, engineering, industrial design, materials, allow architecture to evolve beyond static functionality and become an active participant—with the capacity to perceive, react to, and connect—with humans and the natural world [4].

Some examples of contemporary nature-based architecture that adapts to the environment according to bioclimatic principles and are: Agbar Tower, by Jean Nouvel, in Barcelona [5], Museo Soumaya, by Fernando Romero Enterprise, in Mexico City [6], Al Bahar Towers, by Aedas, Abu Dhabi [7], Gherkin Tower, by Norman Foster, London [8].

However, with a glance at the past, human adaptation to nature can assume an even deeper significance, rooting the migration and developments of entire civilisations to the ground.

To go further, the following chapter shows two peculiar cases of 'adaptive architecture'.

2 *Cueva* and *Tomas De Terreno*

The *cueva* (in Spanish) can be considered an example of a "adaptive architecture". The term refers to artificial cavities, specially dug into the hillside.

They are mostly inhabited by nomadic families and can be considered real cave dwellings, due to their climatic characteristics, their strategic position and the concept of social sharing that characterises them. They also represent a natural refuge for people expelled or emigrated from their cities, highlighting their character of permanent transience.

For example, in southern Spain, the syncretism generated by the three main Mediterranean cultures (Catholic, Muslim, and Jewish) and the ancient ways of living and working has led to the construction of a varied architectural heritage, expressed in both urban and agricultural landscapes. The *cueva* takes the form of its own architectural solutions, realised with building materials derived from the surrounding natural environment [9].

This habitat, typical of a society deeply rooted in the land, the result of the combination of various cultures and historical periods, presents a remarkable richness in its forms, taking advantage of and adapting to a peculiar physical environment, and therefore deserves to be understood as a cultural landscape to be preserved over time [10].

Totally integrated into the natural environment, the *cueva* uses clays, silts, chalks, and conglomerates that are easily excavated. Its layout is adapted to the sloping topography of the hillsides, using the valleys as access routes, and excavating the cave in the hillside.

Moreover, these dwellings suffer less from the extremes of the climate, both the winter cold and the summer heat, maintaining a constant temperature of around 19 °C throughout the year.

The *cueva* is an unfinished, organic construction that can be expanded by digging new rooms, capable of adapting to the geological characteristics of the terrain. The terrain is a cultural space, with historically and socially relevant values, which depends not only on its intrinsic value, nor on the objective recognition of experts, but first and foremost on its social acceptance [11] (Fig. 1).

Another form of adaptive architecture can be the phenomenon of the *tomas de terreno* (in Spanish). *Tomas de terreno* is defined as the temporary action of illegally occupying and inhabiting land, "the possession of a piece of land without sale or title" by building a house, hut or shack on it. It is often a response to the lack of housing for citizens who have no other options [12].

In Latin America in the 1960s and 1970s, land appropriation was understood as a result of marginalisation processes associated with industrialisation and modernisation. In the 1970s and 1980s, in the convulsive social and political context experienced in the subcontinent, land occupations were understood as places of resistance, as spaces of great political connotation and material expression of class struggle [13].

An example of this is the case of the urban social movements, and specifically the land appropriations of the 1973–1985 period in Santiago de Chile, which carried out forms of settlement on these lands, expressing the dual need for permanent housing and resistance to the repression exercised by the military dictatorship during that period [14].

In the 1990s, in a context marked by the restoration of neoliberal democracies, *tomas de terreno* is conceptualised as "encampments" or "irregular settlements", giving rise to a housing policy that sought to reduce the housing deficit [13].

Today, *tomas de terreno* is a strategy used by citizens to reduce residential segregation, facilitate access to social housing and choose a better location within the city, forming a response to the supply of social housing that is usually peripheral and far from the social circuits themselves, as well as from the opportunities offered by urban spaces [13].

Tomas de terreno is an informal habitat understood as a complex socio-spatial reality and a transitional and adaptive housing solution, but at the same time permanent, through a natural procedure of city construction, according to social and spatial processes that have influenced the occupation, construction, transformation and consolidation of these lands, determining particular modus of appropriation—or re-appropriation—that has defined a true informal and spontaneous architectural model [12].

Land appropriation as a subversive form to the state of exclusion is the response to a great demand: the right to a home in a democratic city [15] (Fig. 2).

Fig. 1 Facade of a *cueva* in Fonelas in the province of Granada (Spain) [9]

Fig. 2 *Tomas de terreno* in the ravines in the province of Valparaíso (Chile) [12]

References

1. Merriam-Webster Dictionary. https://www.merriam-webster.com/dictionary/adaptation. Accessed 24 Nov 2022
2. Enciplopedia Treccani. https://www.treccani.it/enciclopedia/adattamento. Accessed 24 Nov 2022
3. CRC—Collaborative Research Center 1244, Adaptive skins and structures for the built environment of tomorrow. https://www.sfb1244.uni-stuttgart.de/en. Accessed 24 Nov 2022
4. Fox, M.: Interactive Architecture—Adaptive World. Princeton Architectural Press, New York (2016)
5. Agbar Tower. http://www.jeannouvel.com/en/projects/tour-agbar. Accessed 25 Nov 2022
6. Museo Soumaya. https://fr-ee.org/es/museo-soumaya-proyecto. Accessed 25 Nov 2022
7. Al Bahar Towers. https://en.wikiarquitectura.com/building/al-bahar-towers. Accessed 25 Nov 2022
8. Gherkin Tower. https://www.fosterandpartners.com/projects/30-st-mary-axe. Accessed 25 Nov 2022
9. Navarro, F.A., Javier, S.: Arquitectura etnográfica en las comarcas de Guadix-Baza (Granada). Necesidad de constituir paisajes culturales. Gazeta de Antropología, (2009). https://doi.org/10.30827/Digibug.6905
10. Urdiales, M.E.: La Cueva: ¿vivienda marginal? Análisis en Benalúa de Guadix. Cuadernos geográficos de la Universidad de Granada, pp. 165–196 (1986)
11. Ortega, J.: El patrimonio territorial: El territorio como recurso cultural y económico. **1**, 31–48 (1998). https://doi.org/10.24197/ciudades.04.1998.31-48
12. Pino Vásquez, A., Ojeda Ledesma, L.: Ciudad y hábitat informal: Las tomas de terreno y la autoconstrucción en las quebradas de Valparaíso. Revista INVI **28**, 109–140 (2013). https://doi.org/10.4067/S0718-83582013000200004
13. De Armas-Pedraza, T., Gascón-Martín, F., Muñoz-Salazar, P.: Percepción de riesgos socioambientales en tomas de terreno de Playa Ancha, Valparaíso Chile. Los casos de Pueblo Hundido y Vista al Mar. Pap Pob **23**, 181–206 (2017). https://doi.org/10.22185/24487147.2017.93.026
14. Becerra, R.I.R., Valencia, M., Hevia, R.O.: La ciudad invisible: Tomas de terreno en Santiago de Chile. 1973–1985. Un escrito revisitado. DU & P: Revista de Diseño Urbano y Paisaje 30–42 (2016)
15. Valenzuela Marchant, C.: El movimiento de pobladores en Chile y las tomas de terrenos como principal sujeto territorial en la segunda mitad del siglo XX. The Movement of Settlers and their Land Occupations as the Main Chilean Territorial Subject in the Second Half of the 20th Century (2020). https://doi.org/10.5281/ZENODO.3928754

Marco Iannantuono is a Ph.D. candidate in Architecture and Design Cultures at Alma Mater Studiorum University of Bologna (XXXVI Cycle). His Ph.D. research, entitled "Methods and tools for digital optimisation of building technologies", is based on the integration of different technologies to achieve a multi-benefit approach in existing building renovations using digital methods and tools.

Temporary Uses/Pop-Up Spaces

Giulia Marzani

Abstract

The traditional way of planning the city is often too slow to meet the quick changes in the needs and behaviour of inhabitants, which are occurring much faster than in the past. Cities' layouts change accordingly, with the consequence that many buildings and infrastructures remain under-used or empty since they are no longer able to fit the needs of their users. From a spatial planning perspective, abandoned or underused buildings are considered the most valid alternative for cities to thrive without consuming a large amount of virgin soil, a finite resource. However, reuse practices are often at odds with economic, political and technical barriers that may prevent the transformation. Temporary uses can represent a lever in this perspective, giving citizens and communities the opportunity to affirm their right to participate in city life through proximity activism. In addition, the implementation of pop-up environments for temporary developments represents a possible solution to meet a city's needs of flexibility, adaptation, and resilience. Pop-up spaces can be seen as forms of spatial and social innovation, allowing new actors to contribute to urban transformation, giving a voice to groups of people who would otherwise be invisible. Pop-up environments inside empty buildings have allowed citizens to appropriate these spaces for the creation of places for productivity and work, as well as for artistic and cultural events.

Keywords

Temporary uses · Pop-up · Urban regeneration

1 Temporary Uses

Urban strategies have changed throughout history, changing the way cities are planned. Contemporary cities are increasingly subject to transformations. Social needs, citizens' behaviour and innovation in technologies evolve over time and cities' configuration change accordingly, with the consequence that many buildings and infrastructures remain under-used or empty since they are no longer able to fit the needs of their users. The spatial planning approach should be able to provide, among others, strategies and tools for implementing adaptive actions so that cities can adapt to change and create resilient communities [1]. One starting point for examining adaptive actions from the perspective of

G. Marzani (✉)
Department of Architecture, University of Bologna, Bologna, Italy
e-mail: giulia.marzani3@unibo.it

A. Borsari et al. (eds.), *TEMPORARY: Citizenship, Architecture and City*, The City Project 4,
https://doi.org/10.1007/978-3-031-36667-3_19

spatial planning is considering abandoned or under-used buildings as the most valid alternative for cities to thrive without consuming a large amount of virgin soil. It also addresses the global challenge of resource scarcity, supporting one of the main environmental challenges at European level; it contributes to solving the global problem of land degradation and consumption which may impact on the increase of natural disasters and social issues such as migration. Land is indeed a finite resource, and it is consumed due to many human activities such as urbanisation and the creation of new infrastructures that should meet the increasing demand of people to live in urban areas. It is estimated that over 500 km^2 of agriculture or natural land disappear every year in the EU, as it is converted into artificial areas,[1] and both Europe and the United Nations have encouraged Member States to achieve "no net land take" by 2050, meaning that regeneration practices have to be developed, focusing more on the rehabilitation and adaptation of the existing built environment rather than on new constructions. Moreover, adaptive actions are considered of key importance for urban systems in the transition towards the new paradigm of sustainable circular cities [1] that is gaining momentum in the European scenario with many cities embracing the circular economy challenge [2].

However, reuse practices often are at odds with economic, political and technical barriers that may prevent transformation [3] and temporary uses can represent a lever in this perspective; the temporary and reversible nature of the transformation can also help in the celerity of execution and in testing the transformation's effectiveness, bypassing administrative procedures that can be shorter, especially if compared to the demolition and reconstruction process [4].

In addition, the reuse of existing properties through temporary activities offers the opportunity for citizens and communities to affirm their right to participate in city life through proximity activism. Reuse, especially temporary reuse, is also used with the aim to explore the different opportunities that the building or the site can offer by using them in different way and involving different actors such as activists, a government, or by creative practitioners seeking affordable work spaces. Temporary uses can also be seen as an opportunity for users to show alternatives to demolition, by protesting against pending demolition acts.

Reuse and temporary reuse offer the possibility of transforming and reinventing the city according to specific desires and through forms of social and civic activism. The temporary status of migrants often limits their mobility and restrains their access to public spaces, also influencing the way they live urban spaces [5]. These challenges aside, migration can be seen as a driver for regeneration processes, especially in areas that suffer from population decline like the rural ones [6]. In the framework of RURITAGE project,[2] it is possible to find an example of how the adaptive reuse of buildings has enabled migrants to join the local community, fostering integration and the creation of an inclusive society. The "role-model" identified by the project is the initiative undertaken by PIAM Onlus, an NGO working in Italy's Asti province to integrate migrants into society. In 2014, the Prefecture of Asti requested an increasing number of hospitality structures for migrants in order to tackle an emergency situation. PIAM Onlus, in cooperation with the COALA Consortium, obtained permission to host foreigners in the historic building of Villa Quaglina, an ancient villa owned by Oblates of St. Joseph, located in Asti countryside and semi-abandoned for some years. Migrants have been involved in the restoration works and the renovation has made it possible to host 45 migrants, with the creation of guest rooms and more flexible spaces for the organisation of various events such as conferences, film screenings and parties. Regeneration practices in the local area include other heritage-based activities which have seen migrants fully

[1] European Commission, https://ec.europa.eu/environment/soil/index_en.htm.

[2] The European project "RURITAGE: Rural regeneration through systemic heritage-led strategies" has received funding from the European Union's Horizon 2020 research and innovation programme under grant agreement no. 776465. www.ruritage.eu.

integrate into the local community. The cultivation of ancient grains and distribution across the territory have fostered relationships with local farmers, while the recovery of an old vineyard and its cultivation has involved a collaboration with oenologists who have trained migrants for this purpose. This has led to the creation of a migrant-ethnic cuisine catering initiative that is still ongoing in the province.

2 Pop-Up Spaces

Besides the needs to plan future urban scenarios and enact building and technical regulations in order to improve citizens' quality of life and urban quality, technological innovation now affects the way people live within the city, with the emergence of new infrastructure, new land use and new space availability. The traditional way of urban planning is often too slow to meet the quick changes in the needs and behaviour of inhabitants, which are occurring much faster than in the past. In this context, the implementation of pop-up environments for temporary developments could be a solution to meet a city's needs of flexibility, adaptation, and resilience [7].

A pop-up environment can be intended as a type of construction that due to characteristics such as lightweight technologies and fast and easy assembly/disassembly operations, occupies the ground only temporarily, being adaptable to different uses and target groups thanks to its flexible structures. Reversibility, adaptive reuse and recyclability of units and components are key criteria for the transition from a linear model of consumption and production to the circular economy one. Focusing on this latter, its basic principles consist in designing out the concept of waste and keeping materials and resources in the economic system as long as possible and at their higher value. If these principles are applied to adaptive reuse and recyclability of components, there are advantages in terms of celerity and affordability that would not normally be achieved with a traditional permanent design, whilst reducing construction waste [8]. In this respect, pop-up environments are associated to circular economy principles and many cities have experienced them on their pathways to reaching the status of circular cities, since they represent a demonstration of how cities can improve their adaptive capacity fostering reuse and recycling opportunities [1]. Pop-up places demonstrate the flexibility of urban sites, suggesting they can be continuously transformed by different temporary uses. In general, pop-ups produce removable and mobile places, able to relocate in a short time and adapt to new sites [9]. The PLACE/Ladywell pop-up is a concrete example of the application of the London Circular Economy route map in the construction sector. Lewisham Council, with support of the Mayor of London, funded a project that enabled the reuse of the former Ladywell Leisure Centre, which was demolished in 2014 and left vacant. The project consisted of creating buildings in which the upper floors are temporary houses managed by Lewisham Council while the ground floors offer affordable mixed-use spaces for local start-ups and entrepreneurs. The temporary housing development will remain on site for between 1 and 4 years, providing 24 homes for local people, thereby contributing to solving the housing crisis in London. In addition, all units exceed the current standard space requirements by 10%, helping the council to meet an existing shortfall in high quality temporary and two-bed accommodation while long term regeneration plans are being developed. PLACE/Ladywell has been constructed with prefabricated and modular cells that can be dismantled and moved to another location within the borough after several years on site.

Pop-up spaces are experimental sites and can be seen as forms of spatial and social innovation, allowing new actors to contribute to urban transformation. In this respect, temporary transformation can give a voice to groups of people who would otherwise be invisible, even if engaging the social dimension with this kind of temporary use is not easy to achieve. However, it can be seen as a driver for social inclusion since pop-up spaces often show the real needs of communities, triggering the public debate and allowing the people involved in the temporary use project to become or return to be part of society. In addition, volunteers that might participate in the projects can become social innovators by committing their time and energy [10]. Moreover, the positive impacts that pop-up environments have brought to cities can be seen from various perspectives. First, the uniqueness of the place, a social-spatial innovation with the interesting combination between the design and the actors that managed to create it. Second, the benefits such as social cohesion and the place-related economic impact generated with the establishment of new hubs and leisure functions. Third, the difference and variation that these types of places would bring to the urban environment, providing new possibilities, dynamism and controversy which are valuable for the city [11]. Pop-up environments inside empty buildings have allowed citizens to appropriate these spaces for the creation of productivity and work places, as well as for artistic and cultural events. In addition, not often are pop-up spaces born in vacant spaces but the way in which pop-up identifies vacancy relies upon a normative approach. Indeed, spaces of transgression or spaces for marginalised social groups are not recognised when spaces are designated as empty and are ready for pop-up experiments.

References

1. Williams, J.: Circular Cities, a Revolution in Urban Sustainability. Routledge (2021)
2. Foster, G., Saleh, R.: The adaptive reuse of cultural heritage in European circular city plans: a systematic review. Sustainability **13**, 2889 (2021)
3. Yung, E.H., Chan, E.H.: Implementation challenges to the adaptive reuse of heritage buildings: towards the goals of sustainable, low carbon cities. Habitat Int. **36**(3), 352–361 (2012)
4. Viola, S., Diano, D.: Repurposing the built environment: emerging challenges and key entry points for future research. Sustainability **11**(17), 4669 (2019)
5. Tan, Y.: Temporary migrants and public space: a case study of Dongguan, China. J. Ethn. Migr. Stud. 1–17 (2020)
6. Conticelli, E., De Luca, C., Egusquiza, A., Santangelo, A., Tondelli, S.: Inclusion of migrants for rural regeneration through cultural and natural heritage valorization. In: Gargiulo, C., Zoppi, C. (eds.) Planning, Nature and Ecosystem Services, pp. 323–332. FedOAPress, Naples (2019)
7. Bertino, G., Fischer, T., Puhr, G., Langergraber, G., Österreicher, D.: Framework conditions and strategies for pop-up environments in urban planning. Sustainability **11**, 7204 (2019)
8. Ellen Mac Arthur Foundation: Cities in the Circular Economy: an Initial Exploration (2017)
9. Harris, E.: Navigating pop-up geographies: urban space–times of flexibility, interstitiality and immersion. Geogr. Compass **9**(11), 592–603 (2015)
10. Urban Agenda Partnership on Circular Economy: Sustainable & Circular Re-use of Spaces & Buildings —Handbook (2019)
11. Lehtovuori, P., Ruoppila, S.: Temporary uses producing difference in contemporary urbanism. In: Transience and Permanence in Urban Development, pp. 47–63 (2017)

Giulia Marzani is a Ph.D. candidate in Architecture and Design Cultures at Alma Mater Studiorum University of Bologna (XXXVI Cycle). Her research project, entitled "Circular cities. A decision-making framework in support of innovation and urban planning", aims at systematizing the current knowledge about the application of circular economy principles in cities with specific emphasis on urban planning.

Classroom/Digital Citizenship/Material Culture

Angelo Massafra

Abstract

Classrooms are emblematic spaces of educational buildings. In classrooms, students have lessons and exams, both activities related to the principles of collectivity and institutionality. For this reason, the classroom is a crucial place for forming students' citizenship where the whole community welcomes them. In classrooms, off-site students, who are temporary actors in cities, interact daily with society and its institutions. The unlimited flexibility that is potentially typical of the classroom object, intended in its tectonic meaning, is usually not registered in heritage buildings. Its staticity is incompatible with new learning methods. Therefore, a rethinking of classrooms is expected to allow new relationships between temporary subjects and the community. /Virtual spaces are more than ever connected with the real world, allowing new interactions between cities, urban services and citizens. Consequently, new relationships between human beings and the virtual world have been born in recent times, and more attention is paid to investigating the splitting between the real and the digital identity of individuals. Starting from these themes, the concept of digital citizenship has risen. It refers to the individual's capability to consciously use virtual communication tools to take advantage of networked services. The potential is clear, but there are many complexities and risks, as digital tools may be limited in terms of accessibility. The digital divide is one of the most evident results of the ongoing digitisation, tending to exclude some temporary categories of the population from the digital, producing considerable socio-economic and cultural damage. /Material culture refers to that portion of the anthropic environment made or changed by society, consciously or unconsciously, depending on cultural and productive relationships. In the architecture and construction domains, analysing the material culture at the urban level implies searching for the links between the built environment (intended as a complex set of human products) and the social issues for which it was materially shaped in a given historical period. This approach is significant for ancient and modern ages and for structured and temporary societies. Indeed, transient phenomena may stand out sharply by emerging from the best-known and most accessible contexts, although these may not be well documented, and their in-depth analysis could help discover unknown aspects.

A. Massafra (✉)
Department of Architecture, University of Bologna, Bologna, Italy
e-mail: angelo.massafra2@unibo.it

A. Borsari et al. (eds.), *TEMPORARY: Citizenship, Architecture and City*, The City Project 4,
https://doi.org/10.1007/978-3-031-36667-3_20

Keywords

Classroom · Educational buildings · Temporary students · Digital citizenship · Digital identity · Digital divide · Material culture · Construction types · Urban transformations

1 Classroom

1.1 Classroom in Built Heritage

Many buildings were constructed between the Twenties and the Sixties of the twentieth century within the Italian-built fabric, the so-called "Modern Age". Among these, there are numerous university public buildings characterised by substantial aesthetic, historical, and architectural values and by particular relationships with the urban context.

These architectures preserve identity significances of common interest, so they must be maintained in terms of socio-cultural recognition and for their functional role played within the urban apparatus. At the same time, this category of buildings is affected by some characteristics genetically intrinsic to its construction type. At present, they usually show an apparent incompatibility with the social, environmental and economic needs that they are required to satisfy daily, with the conditions of use and enjoyment of their users, and with the sustainability issues of their management [1]. For instance, modern university buildings are often poor in terms of functional space flexibility, energy efficiency, resource consumption, operational costs, greenhouse gas emissions, smart readiness, and, above all, user safety and comfort (both in terms of thermal, visual, and acoustic comfort and of amenities availability). A symbolic case is the building of the Faculty of Engineering in Bologna, designed by Giuseppe Vaccaro in the 1930s according to a requirement framework very different to the contemporary one. In fact, the building was designed for hosting about 300 students; nowadays, the number of students that every day occupy the buildings is about ten times more [2].

The functions within these buildings are manifold and change over time: representative, administrative, teaching, research, service, and care. The various spaces include offices, classrooms, laboratories, libraries, study rooms, auditoriums, sports facilities, residences, recreational areas, archives, storerooms, and refreshment areas.

Within their dual functional and tectonic meaning, classrooms can be considered emblematic spaces of the current inadequacy of this category of buildings [3]. This inadequacy regards the operational conditions for users, the management practices of their administrators, and the poor performance of sustainability and environmental quality that they demonstrated compared to what imposed by current regulatory standards.

1.2 Citizenship in the Classroom

Please note that the first paragraph of a section or subsection is not indented. The first paragraphs that follows a table, figure, equation etc. does not have an indent, either.

The term "classroom" can be understood both from the functional point of view—as a "room intended for important and solemn meetings, such as those of courts and parliaments, or for lessons in schools and universities" [4]—and from an architectural point of view—as an ample free space capable of offering infinite flexibility for carrying out various activities. The classroom usually hosts activities intrinsically linked to the principles of collectivity and institutionality, such as lessons and exams. For this reason, the classroom is a key place for the implementation of a constant and daily implicit act of citizenship by its users, who recognise roles and exercises within the classroom (e.g., professors and students), actively and ordinarily exercising rights (e.g., possibility of enjoying the lesson) and duties (e.g., rules of good conduct towards other auditors).

1.3 Rethinking Classrooms

With regard to off-site students, who are temporary actors in cities, the classroom can be intended as a privileged place where individuals can develop cognitive processes of interaction and integration with society on a daily basis, which in this case is recognised in the university community and the city. Therefore, the classroom is a place of opening the building to citizenship, where temporary city players such as students are welcomed. The cognitive interaction [5], which is woven between society and its temporary subjects within the classroom, coincides with the involvement of these subjects within a social group and the conversion of these figures from simple transient users of the building to conscious community members.

The temporary nature of these users needs to be reflected in how they are connected in classrooms. For instance, if the learning activity can be considered a permanent relationship in classrooms, the same cannot be said for the learning methods. From this point of view, the potential unlimited flexibility that is tectonically typical of the classroom object is not reflected in the classrooms present in most heritage buildings [6]. Instead, these existing spaces are characterised by a tension between the static rigidity of their physical conformation and the mutability of the situations and needs that they host.

The increasingly frequent use of learning methods based on interactive approaches and the exponential growth of e-learning, [7] fuelled by the pandemic, have made clear the need for a physical and technological rethinking of teaching spaces: a rethinking that does not simply provide prescriptive solutions in which the physical environment is detached from its users, but that could create plural environments supporting learning and the active participation of both students and teachers, redefining the role of the physical space in the process of interaction between the temporary subjects and the community.

2 Digital Citizenship

2.1 The Digital Twin Paradigm

In the last decades, the rapid pace of digitisation has produced remarkable advancements in new technologies for acquiring, organising and processing information. Nowadays, digital transition seems a moment to be seized in order to materialise effective technical, economic and social improvements for society in various areas.

In the Architecture, Engineering and Construction sector (AEC), the emerging paradigm of the digital twin, [8] defined in the manufacturing field in the early 2000s, is opening the door to the world's virtual dimension. This dimension can be imagined as a futuristic scenario where the virtual space will continuously dialogue with the real world through a bidirectional interaction between cities, urban services and citizens. For instance, some researchers have started investigating whether adopting these cyber-physical systems in existing buildings could directly involve building users in building management by collecting their feedback [9]. In this way, the users' qualitative digital "testimony" is considered at the same level as the quantitative "testimony" from the sensors for monitoring the conditions of use and environmental quality of indoor and outdoor spaces, and thus the health and well-being of workers [10].

2.2 Digital Identity and Digital Citizenship

The theme of digitisation is not merely technical but also refers to the social sphere. The exponential spread of new technologies and social media has caused people to reflect on the relationships between human beings and the virtual world in recent periods. The "digital identity" is one of the issues most addressed by sociologists and psychologists in the twenty-first century. Much attention is paid to the splitting between the real and the digital identity of individuals. Particular relevance in the debate is assumed by the connection between the individual and the emerging digital society, a theme that has given rise to the concept of digital citizenship [11–13].

In its traditional conception, citizenship is defined as the bond or condition of being related to a state by rights and duties that such a relationship entails [14]. It results from three connections: citizen-government, citizen-public services, and citizen-community. On the other hand, digital citizenship represents the individual's capability to make aware use of virtual communication tools to take advantage of networked services and develop critical thinking and consciousness about the possible risks and opportunities of surfing the Internet. While citizenship usually refers to the concept of belonging to a state, digital citizenship includes relationships with a broader community, which is online and virtual.

Even in the virtual world, there is a divarication within this panorama between the two prevalent meanings of citizenship: the legal/formal and the social/substantial [13]. The first one pertains to the relationship between the citizen and the state, including issues such as the possibility of accessing online services, obtaining certificates, interacting with public administrations, making payments, and consequently the introduction to new tools such as digital identity and the citizen's digital domicile. The second relates to the relationship between citizens themselves, inducing the possibility of having work and study, and the cultural or leisure relationships within the community network beyond the state's geographic boundaries. It requires skills related to communication, content use and production, copyright, and knowledge of the many issues that take new forms on the web.

2.3 The Digital Divide

The complexities and potential brought by digital processes are many. Hence, digital tools play a predominant role [14] in terms of accessibility. At the basis of the digital transition of the community, there is the need to access the Internet, intended as a top right that enables the exercise of other fundamental rights, such as the freedom of expression, the freedom of informing and being informed, the economic initiative, and the possibility to innovate. Therefore, consciously accessing the Internet has become a necessary condition for complete individual and social development, as indicated by the Declaration of Rights on the Internet and sanctioned by numerous supranational organisations (including the UN, G8 and EU).

The issue of accessibility, or rather its rejection, is nowadays a crucial aspect of social vulnerability. The so-called "digital divide" is one of the most evident results of the ongoing process of digitisation. It describes the adverse effects of technological progress and innovation that tend to exclude from digital—by choice or by chance—specific transitory categories of the population (e.g., migrants), who lose the benefits of digital, producing considerable socio-economic and cultural damage [15].

3 Material Culture

3.1 Definition

Material culture represents that part of humankind's physical world that has been created or transformed consciously or unconsciously by groups of people according to culturally dictated plans [16, 17]. In archaeology, the term refers to the set of physical products of a society, i.e. the objects made or changed by humans [18]. These objects surround people and their activities and have specific physical, chemical, biological and morphological properties that result from the circumstances in and for which they were made. Since objects do not sit alone in the human or natural world, when they interact with humans or other human products, they gain meaning and develop relational characteristics [19], which explicit their capabilities to interact with the social issues of the real world depending on the geographical and historical context at a given time.

Understanding the material culture of past and contemporary societies means comprehending the relationships between the complex set of actions carried out by human groups within society and the material and physical products that result from these actions. In other words, the knowledge of material culture implies studying the connections between the artefacts and the technical, social and economic behaviours and procedures enacted to produce, exchange, use, break down, discard and abandon physical products. It considers all possible "actions and retroactions" between the different production sectors around which the group is structured, as well as a function of the time variable following this approach [18]. In this sense, investigating the artefacts produced by a specific human group and their meanings through a multidisciplinary and integrated method could allow to discover non-material aspects and learn about the needs, knowledge and intentions behind past or recent societal choices, changes and modifications [20].

3.2 Urban Material Culture

The archaeological definition of material culture can be extended to the fields of study of Architecture, Built Environment and Construction. For instance, analysing material culture at the urban level means searching for the associations between specific physical categories of the built environment and the social and urban issues for which they were built in a given historical period. On the one hand, examining urban material culture implies delving into the construction logic of existing buildings and districts, intended as sets of anthropic artefacts. On the other hand, it consists of understanding the social, technical and economic needs that physically shaped the built environment and, therefore, the society's production apparatus.

An analysis of the material culture of the architectural heritage of a specific period thus displays not merely the buildings as products of an isolated production system, but also the relationships between these products and the whole production chain of that time. More or less evident, these linkages could be identifiable as physical transformations of the urban apparatus, which depend on common construction processes and common morphological types.

Following the thread of urban material culture suggests searching for examples of physical manifestations of economic and social phenomena that allow an interpretation of the transformations of construction and production systems, and thus societal behaviours.

This statement is valid both for ancient and modern epochs, and for structured and temporary societies. While the knowledge of the permanent activities in a given geographic context is usually facilitated by a greater availability of documentary sources (linked to major events and large masses), temporary phenomena may stand out sharply by emerging from the best-known and most accessible contexts, although they may not be well documented.

Identifying the proper category of artefacts to be analysed as manifestations of the variation of urban form in a specific period can help perceive the past temporal instances of a given social structure or part of it, which could be equally temporary to urban modifications.

3.3 Modern Material Culture

A significant example of modern material culture can be retrieved by observing the fragmented apparatus of the suburban neighbourhoods built in Italy after World War II. These segments of Italian cities physically reflect the chaotic atmosphere of the Reconstruction period, as well as the poor development of the national industry sector at that time. The buildings that rose in such areas between the 1950s and the 1960s, which amount to a large percentage of Italy's built heritage, were constructed mainly according to the INA-Casa plans [21]. These plans were designed to give housing to an artisanal society transitioning from the destruction brought by the war to the industrialisation of the country's economic and productive machine. Therefore, the small enterprises spread widely throughout the Italian territory were fostered, triggering a mechanism whereby unoccupied and unspecialised people were directly called upon to work to build their houses, producing primary goods that they themselves would enjoy [22]. This resulted in poorly controlled construction sites and non-standardised construction procedures [23], therefore the adoption of a particular construction type, the so-called "mixed construction".

Another case comes from the built heritage of the University of Bologna, which has continuously developed in parallel with the transformation of the city's urban fabric since the University's foundation in the Middle Ages. In the twentieth century, migration phenomena greatly influenced the city's demographic transformations and was the leading cause of the intense urbanisation. The changes dictated by the Athenaeum in that period allowed to increase the capacity of teaching spaces in the city, satisfying the exponential demographic development of those years [24]. This was directly reflected in the physical shape of the town that underwent the effects of the post-war urban sprawl, involving the renovation of historic buildings and new constructions.

References

1. Massafra, A., Gulli, R.: Strumenti e modelli per la gestione digitale del patrimonio costruito. Tools and methods for the digital management of built heritage. In: Colloqui, A.T. (ed.) 2022—Memoria e Innovazione, pp. 1241–1260
2. Gulli, R., Predari, G.: Il Moderno a Bologna: la Facoltà di Ingegneria di Giuseppe Vaccaro (1932–35). In Ananke **83**, 101–108 (2018)
3. Barret, P., Treves, A., Shmis, T., Ambasz, D., Ustinova, M.: The Impact of School Infrastructure on Learning. A Synthesis of the Evidence. International Bank for Reconstruction and Development (2019)
4. Treccani: Dizionario della lingua Italiana. Istituto della Enciclopedia Italiana. Roma (2017)
5. Weinstein, Y., Sumeracki, M.A.: Understanding How We Learn: A Visual Guide. David Fulton, London (2019)
6. Borri, S.: The Classroom has Broken. Changing School Architecture in Europe and Across the World. INDIRE, Florence (2018)
7. Kaplan, A.M., Haenlein, M.: Higher education and the digital revolution: about MOOCs, SPOCs, social media, and the Cookie Monster. Bus. Horiz. **59**(4), 441–450 (2016). https://doi.org/10.1016/j.bushor.2016.03.008
8. Grieves, M., Vickers, J.: Digital twin: mitigating unpredictable, undesirable emergent behavior in complex systems. In: Kahlen, J., Flumerfelt, S., Alves, A. (eds.) Transdisciplinary Perspectives on Complex Systems. pp. 85–113. Springer International Publishing, Cham (2017). https://doi.org/10.1007/978-3-319-38756-7_4
9. White, G., Zink, A., Codecà, L., Clarke, S.: A digital twin smart city for citizen feedback. Cities **110**, 103064 (2021). https://doi.org/10.1016/j.cities.2020.103064
10. Jayathissa, P., Quintana, M., Abdelrahman, M., Miller, C.: Humans as a sensor for buildings: intensive longitudinal indoor comfort models. Buildings **10**(10), 174 (2020)
11. Hintz, A., Dencik, L., Wahl-Jorgensen, K.: Digital Citizenship in a Datafied Society. Polity Press, Cambridge (2019)
12. Mosseberger, K., Tolbert, C., McNeal, R.: Digital Citizenship: The Internet, Society, and Participation (2007)
13. Zook, C.: What Is Digital Citizenship & How Do You Teach It? https://www.aeseducation.com/blog/what-is-digital-citizenship. Accessed 06 May 2022
14. Mossberger, K., Tolbert, C.J., Stansbury, M.: Virtual Inequality. Beyond the Digital Divide. Georgetown University Press, Washington, D.C. (2003)
15. Connected Education for Refugees: Addressing the Digital Divide. https://www.unhcr.org/publications/brochures/61b743ef4/connected-education-refugees-addressing-digital-divide.html. Accessed 21 Nov 2022
16. Smelser, N.J., Baltes, P.B.: International Encyclopedia of the Social & Behavioral Sciences. Elsevier, New York (2001)
17. Woodward, I.: Understanding Material Culture. Sage. Los Angeles, London, New Delhi, Singapore (2007)
18. Mannoni, T., Giannichedda, E.: Archeologia della Produzione. Einaudi, Turin (2011)
19. Schiffer, M.: Material Life of Human Beings: Artifacts, Behavior and Communication. Routledge, London (1999)
20. Burchardt, M., Hohne S.: The infrastructures of diversity: materiality and culture in urban space—an introduction. New Diversities **17**(2), 1–13
21. Bardelli, P.G., Capomolla, R., Vittorini, R.: L'architettura INA Casa (1949–1963): aspetti e problemi di con-servazione e recupero. Gangemi, Rome (2003)
22. Di Biagi, P.: La grande ricostruzione: il piano Ina-Casa e l'Italia degli anni Cinquanta. Donzelli, Rome (2010)
23. Poretti, S.: Modernismi Italiani. Architettura e Costruzione del Novecento. Gangemi, Rome (2008)
24. Gulli, L.: Politiche territoriali e sviluppo dei poli universitari nella regione Emilia-Romagna (1945–1968). In: E-Review, vol. 5, (2018)

Angelo Massafra is a PhD Student in Architecture and Design Cultures at Alma Mater Studiorum University of Bologna (XXXVI Cycle). His PhD research, entitled "Topology-oriented Digital Twins for the Performance-based Management of Built Heritage", aims at developing a digital twin platform prototype for enabling the performance-based management of existing heritage buildings.

Grenze Versus *Schranke*/Open Form

Claudia Nigrelli

Abstract

The first part of the paper discusses the gradual emergence of the distinction between the German terms *Schranke* and *Grenze* in the history of European philosophy. Where *Grenze* tends to denote a distinction-relation between two terms, *Schranke* could be summarized as an absolute negation. These terms are discussed through Kant, Graham Priest, Hegel, Josef König, Helmuth Plessner, Frederik Buytendijk, and Richard Sennett. The second part of the paper discusses the concept of open form: boundlessness, free order, fluidity, and movement seem to be some of its characteristics. The discussion is developed from a comparison of Heinrich Wölfflin's art history and Helmuth Plessner's theory of the organisms and philosophical anthropology. Finally, the author examines the use of the concept of open form by Umberto Eco, Bruno Zevi and Richard Sennett.

Keywords

Boundary · Limits · Antinomy · Form · Open form

C. Nigrelli (✉)
Department of Architecture, University of Bologna, Bologna, Italy
e-mail: claudia.nigrelli2@unibo.it

1 Grenze Schranke

According to the *Historisches Wörterbuch der Philosophie* (1974), both expressions enter the vocabulary of German philosophy through Leibniz and the German translation of the terms *limes* and *terminus* during the eighteenth century. *Schranke* is intended to be a translation of *limes*, whose relevance concerns the limits of a finite thing [1], whereas in Leibniz the *Grenze-terminus* relates more directly to infinitesimal calculus and is therefore, in essence, a mathematical concept. While in the eighteenth century the two terms were sometimes used as synonyms, with Leibniz, but especially with Kant, they definitively differentiated one another. *Schranke* indicates in Kant a negation, a lack, a closure. *Grenze*, on the other hand, indicates something positive, implying a space of ulteriority with respect to what it delimits or encloses. In *Grenze*, something acquires determinacy and completeness because it is defined with respect to something else, a double distinction/relation of the type 'limited/unlimited', 'possibility/impossibility'. *Grenze* is *Grenzlinie*, a line-boundary in which qualitatively different fields meet, a "participatory conjunction" [2] in which the boundary belongs to the two regions it divides, bringing them into relation. If *Schranke* thus indicates an insurmountable "division" or "barrier", the notion of *Grenze* is instead invested with a double negative and positive meaning: negation of continuity, on the

A. Borsari et al. (eds.), *TEMPORARY: Citizenship, Architecture and City*, The City Project 4,
https://doi.org/10.1007/978-3-031-36667-3_21

one hand, but also recognition of this discontinuity that makes the relation dynamic. The Kantian *Grenze* as the nexus between reason and something completely unknown becomes the constitutive sign of all human enquiry and the norm of enquiry itself [3], and yet its positivity lies not only in its cognitive/rational or pragmatic/anthropological valence, but in enabling the relationship between delimited spaces that define themselves reciprocally, precisely from *Grenze*.

In the book *Beyond the Limits of Thought* (1995), Graham Priest discusses what he calls "the limits of thought" (that is: the limit of what can be expressed; the limit of what can be described or conceived; the limit of what can be known; the limit of iteration of some operation or other, the infinite in its mathematical sense) as boundaries beyond which certain conceptual processes cannot go, which are dialetheic, that is, they are locus of true contradiction [4]. The contradiction is that the processes in question do cross these boundaries. Between Closure and Transcendence there is an object that is generated by an appropriate operation both within and without a totality (of all things expressible, describable, etc.). The object is then itself contradictory, as Priest shows through discussing the history of philosophy and logic: from the arising of the question concerning the "limits of thought" in pre-Kantian philosophy, to its theorisation by Kant and Hegel and its formulation in the paradoxes of self-reference and in the modern philosophy of language.

The classical formulation of the problem was given by Kant. I will briefly discuss the example that Priest gives in order to understand Kant's formulation, which results in the persistence of contradiction. Let's focus on a non-technical example concerning what Priest calls the Fifth Antinomy. The example given is the thought of x: as Priest says, "given any object or collection of objects, the thought of it is a distinct object (unless the thought is of a self-referential kind). So the contradiction runs as follows […]. Start with any object, say, the *Critique of Pure Reason*, and apply the generator iteratively, to produce, at each stage, the thought of the previous object: the *Critique*; the thought of the *Critique*; the thought of the thought of the *Critique*. And when we have an unbounded sequence of thoughts we next produce the thought of all of them, and then keep going. […] Let this procedure be performed as often as possible […]. Consider the totality of thoughts generated in this way, T. Clearly, T has no last member, since for any object there is a thought of it […]. And *ex hypothesi*, the generator can be no further applied to T. So T cannot be thought of (Transcendence). But you can think of T: you have just done so (Closure). Contradiction" [5]. Kant's claim is that antinomy is inherent in aspects of our reasoning concerning the limits of thought, but he tries to solve this contradiction in two ways that we will not discuss in detail here. It is important to notice the persistence of contradiction, which according to Priest concerns the object itself, that is, the limits of thought. This problem is central to Kant's *Critique of Pure Reason*, which aims to "chart the logical geography of the most general features of thought and […] spell out their limits" [6]. If the proper of reason is to go beyond the bounds [*Grenzen*] of experience, this means that reason has the resources to sustain normative reflection upon its own claim, that is, it is capable of critical self-refection. Reason itself recognises that it is limited in its use of understanding to the conceptual cognition of the sensory world, but that it can and must think beyond this. The distinction between *Grenze* and *Schranke* highlights this contradiction: as Kant says, "Boundaries [*Grenzen*] (in extended things) always presuppose a space that is found outside a certain fixed location, and that encloses that location; limits [*Schranken*] require nothing of the kind, but are mere negations that affect a magnitude insofar as it does not possess absolute completeness. Our reason, however, sees around itself as it were a space for the cognition of things in themselves, although it can never have determinate concepts of those things and is limited to appearances alone" [7]. He continues by saying that while in mathematics and natural sciences human reason recognises limits but not boundaries, metaphysics leads us to the boundaries, which are the natural predisposition of our reason.

Hegel discusses and develops Kant's distinction between *Grenze* and *Schranke*. The concept of *Grenze* appears in relation to the discussion on how something can be said to be so and so, that is, on 'determinateness' [*Bestimmtheit*]. *Grenzen* are the bounds of the concept of a thing, that is determined by reference to some 'other': as Katerina Deligiorgi remarks, "the concept of a 'bound' captures the 'determination' of something, or what something is, whilst internalising the other relation that frustrated the previous attempt at 'determinateness'. The bound is thus the 'non-being of the other [...]. 'Determination', [Hegel] argues, is 'the unrest of the something in its bound [*Grenze*], in which [the something] is immanent, an unrest which is the contradiction which impels the something out beyond itself'" [8]. *Grenze* is thus what both unites something with its other and separates it from it. This ambivalent function of the bound makes the something and the other as two "ideal" moments of a whole that are qualitatively different. The "something" points beyond itself to its non-being and passes over into it, thus becoming the finite [*Endlich*], which is characterised by the fact that passes away. See the *Historisches Wörterbuch der Philosophie* (1974) p. 876–878 for a further discussion concerning Hegel's reflection on the finite and the infinite in relation to *Grenze* and *Schranke* and for a critique of Fichte's philosophy.

For the latter too, *Grenze* refers to what lies beyond it, but what makes *Grenze* is the self-limitation of the I, which establishes its relation with the world by distinguishing a non-I and by opposing to this non-I. As Hans Heinz Holz remarks, this setting of a "this side" and of a "beyond" of the boundary results in a difference between part and whole. This becomes explicit in Josef König's *Begriff der Intuition* (1926) [9], which Holz sums up as follows: "The visible [*Anschaulich*] is always only a finite section of the infinite, a limited one. The boundary of the limited [*Die Grenze des Begrenzten*] contains the necessary possibility of going beyond it (as a potentiality and as a tendency); it carries in itself the indices 'this side' as well as 'beyond'. This means that it not only separates the limited from the other, but it is also the place of their coming together and thus of the possible transition from the one into the other. It is not a third alongside this and beyond, but the categorical form that constitutes this and that as this and that" [10]. Now, Holz discusses Fichte and König's positions in regard to Helmuth Plessner's concept of *Grenze*, which becomes not only the basic principle of Plessner's philosophy of the organic but also the foundation of philosophical anthropology, precisely because the human being is capable of acknowledging its own limitation and its relation with the world from which it is divided but also connected. Having a boundary, in Plessner's philosophy of the organic, is considered as the "minimal condition" that must be given for there to be life, that is, it concerns the proposition that living bodies appear to us as exhibiting a fundamentally divergent relationship between outer and inner as objective determination. As Plessner says, this means that living bodies have an appearing, intuitive boundary [*Grenze*]: "The boundedness of physical bodies is not to be understood in any derived way, but in its visual and tactile intuitiveness. Outlining and contouring indicate this state of affairs but are not identical with it. While outlines and contours can be drawn, the fact of boundaries can only be understood, not drawn. The factors upon which the boundaries are based and that in physics and chemistry are defined as forces of cohesion, chemical bonding, and so forth, have to be left aside in the logical analysis of this state of affairs" [11]. Plessner's focus on boundaries concerns precisely the philosophical reflection on the relationship of the bounded body to its boundaries. The boundary of living beings is both a contour or spatial property and the place of the reversal of the two directions (inner and outer), which cannot be converted: "the organic shape boundary must, then, as a *gestalt*, have a nature that is beyond *gestalt*, that is not exhausted by *gestalt*" [12]. This is possible because, according to Plessner, organic boundaries not only ensure the transition from the body to the media but they execute this transition and are themselves the transition: the boundary comes into existence here because it is no longer the virtual in-between of the body and the media but "rather of its own accord fundamentally distinguishes between the formation it encloses [*begrenzt*] as such from the other as other. [...] That means that the body's beginning or end is independent of that which exists outside of it, although sensory determination is not in the position to directly demonstrate this independence by means of sensory attributes" [13]. When the body has the boundary, its boundary not only encloses it, but also opens it up to the medium. That means that the body is both within its own boundaries and beyond them; a barrier [*Schranke*], on the contrary, is a restraint that only exercises the function of closing off the area it confines: "*In abstracto*, every barrier [*Schranke*] can also be regarded as a boundary [*Grenze*] provided that it lies in a continuum of possible progression. *In concreto*, however, the difference between boundary and barrier quite often manifests with painful clarity. Essential for the boundary from a point of view situated *behind* it is the unity of closing off and opening up, which from a point of view situated *above* it is the unity of standing still and passing over (from one area *into* another)" [14].

The boundary is thus not a specific entity but the "minimal condition" of the duality of aspects in which the relation between organisms and environments occurs, a duality that is expressed in the way the body appears as *positioning* itself in the environment, and that allows to distinguish the living from the non-living. As J.M. Bernstein remarks, this means that "the living thing is neither a subject, an auto-poietic, self-creating being, nor an object, something made and created by another or merely given, but a subject-object that has both active and passive aspects; indeed, part of what constitutes the category of boundary is its bidirectionality, its implicating two spaces that are not merely adjacent to one another but are in continual interaction with one another, and where the consequence of that interaction is the establishment of one space being inner and the other outer" [15]. A further reflection on the relation between the *Grenze* and the organic form can be found also in Frederik J. J. Buytendijk's *Anschauliche Kennzeichen des Organischen* (1927) [16] and in Georg Simmel's *Lebensanschauung* (1918) [17]. This latter's reflection on *Grenze* also concerns the sociological importance of the notion of boundary. He writes: "The boundary is not a spatial fact with sociological effects, but a sociological reality that is formed spatially" [18]. This notion is thus central to his sociology of space and identified as constitutive of interaction and socialisation. As Gregor Fitzi points out, despite the differences between Simmel and Plessner, "there is [...] a common interest aimed at explaining the connection between human nature, its tendency to produce cultural content and the hermeneutic problem of experience" [19]. Plessner's philosophical reflection concerning the boundaries as the condition of possibility of the organic forms leads to a philosophical anthropology in which boundary-making becomes constitutive of the variety of expressions of the human form of life and thus inherently political [20].

This distinction between a barrier that closes, impermeable, denying what lies outside it, and a boundary that relates two spaces, dividing them, is in place in the terms boundary and border used by Richard Sennett in the wake of Stephen Jay Gould's distinction of two kinds of edges in natural ecology: "Borders are porous edges, boundaries are not. The boundary is an edge where things end, a limit beyond which a particular species must not stray or, conversely, which it guards as do prides of lions or packs of wolves by peeing or pooping to tell others to Keep Out! The boundary marks a low-intensity edge. Whereas the border is an edge where different groups interact; for instance, where the shoreline of a lake meets solid land is an active zone of exchange where organisms find and feed off other organisms. Not surprisingly, it is also at the borderline where the work of natural selection is the most intense" [21]. Now, Sennett applies the same ecological difference to what he calls "human communities", particularly to the urban habitat, which is divided in parts through borders or boundaries. Because of the quality of the border as something that, like cell membrane, must let matter flow in and out, that is, in which porosity exists together with resistance, both protecting and exposing, he points out that "today, the planner's challenge is to create membranes", not necessarily from a material point of view but from a social point of view.

2 Open Form

"Every work of art is something formed, an organism. Its most essential attribute is its characteristic of necessity: nothing can be changed or moved; everything has to be just the way it is" [22]. With these words Heinrich Wölfflin opens the third section of his book *Principles of Art History* (1915), whose title is "Closed and Open Form (tectonic and a-tectonic)". If every work of art has a form that is inevitable, he continues, the way this inevitability is achieved makes the difference: for example, in the Italian Cinquecento a tectonic style was developed to its ultimate perfection, while in the Dutch style of the seventeenth century it was the free, a-tectonic style that existed as "the only possible form of presentment" [23]. The concepts of open and closed form allow to understand this dialectic. The style of the closed form realises, with more or less tectonic means, a self-contained entity, while "the style of the open form always suggests something beyond itself and wants to appear unbounded, even though there will always be a hidden boundary and it is precisely that which makes the closed character possible in the aesthetic sense" [23]. The difference lies in the construction of the form: in the closed form the vertical and the horizontal dominate the picture, there is a central axis or a perfect symmetry between the two halves of the picture and the contents are arranged following a precise order in relation to one another (Wölfflin is discussing the closed form in the sixteenth century). On the contrary, the open form refuses elementary oppositions to predominate, or at least to structure the image (they lose their tectonic force). The order is free: "pure symmetries disappear, or are made inconspicuous by all manner of destabilising devices" [24]. It favours the diagonal instead of the vertical and the horizontal, breaking with the rectangularity of the pictorial space. Instead of showing the necessity of a relationship with the frame, the open composition "ought to look more like a random section of the visible world" [23] that the spectator might enjoy only temporarily, as a passing moment. This type of form eliminates geometry: "living beauty now adheres to unbounded rather than bounded form" [25], in a relation of tensions. Now, the closed and tectonic form is architectural. This is why, according to Wölfflin, the baroque has kept the feeling of pure symmetry as natural within the sphere of architectural form, but has suppressed it in the painting: in the open form, "the significant aspect [...] is not its structure but the breath that animates rigidity with fluidity and movement. Values of being in one case [closed form] and of change in the other [open form]. A beauty of boundedness in the former and of boundlessness in the latter" [26].

Now, this tension between a closed form and an open form is also discussed by Helmuth Plessner. His contribution to the argument concerns firstly the organic form. In his *Levels of the Organic*, Plessner focuses on the two types of form in order to show the relationship between living bodies and their environments. He distinguishes between an open form, that of plants, and a closed form, that of animals and humans. Following Hans Driesch's *The Science and Philosophy of The Organism* (1907) [27], a form is open "if the organism in all of its expressions of life is immediately incorporated into its surroundings and constitutes a non-self-sufficient segment of the life circle corresponding to it" [28]. Plants are thus in direct contact with the medium, their relation between part and whole is less integrated than for animals because each part is capable of growing and reproducing independently, and it remains self-sufficient in relation to the others (grafting, cuttings). The plant does not have any centre tying together or representing the whole body, its individuality is not constitutive but rather an external moment of its form: it is considered a "dividual". In Plessner's view, the form is a concrete living act, that is, it consists in the way the relationship of the organism to its surrounding field is played, it implies an effort and it exposes to dangers. Even if the strict distinction between open and closed form allows Plessner to understand the differences between plants, animals and humans, both openness and closure are always present as a manifestation of the boundary, that is, as an essential index of vitality. That means that "the closed system of the living body corresponds to the closed system of the positional field, but just as the body possesses its own boundaries by at the same time being out and beyond them and opening them, the closed system must too be an open system, allowing for and demanding constant correction" [29].

Now, this tension becomes fundamental in order to grasp Plessner's argument on human beings and the methodology of the Opening Question, which is central to his philosophical anthropology. Because humans are not bound by instinct, they are highly variable historically and culturally or, in other terms, unfathomable to themselves. Their essence is not fixed: "If the human is to be understood as the point of the creative breakthrough of its intellectual world from out of whose values and categories the human understands and deals with itself, its contemporaries, and its environment [...] then the theory that seeks to render this creatorship conceivable can, as we saw, be neither empirical nor a priori. [...] [Rather] it must remain *open*, for the sake of the universality of its view onto human life in the full scope of all cultures and epochs of which the human is capable. This is why the unfathomability of the human moves to the centre of anthropology, and the possibility of being-human that contains what makes the human a human in the first place, that human radical, must yield to the standard of unfathomability. Only if and because we do not know what else the human is still capable of does it make sense to persevere through a life of suffering on this earth. The human's own unfathomability is the binding principle of human life and of human comprehension of life, binding for the sake of the seriousness of its task" [30].

Now, let's go back to Wölfflin. In *Principles of Art History* he affirms that while painting *can* be tectonic, architecture *must* be tectonic. If tectonic is the closed form, that means that architecture cannot be an open form: "for architecture, abolishing the tectonic framework would be tantamount to self-destruction [...] Architecture is at root tectonic and only the decoration seems able to move more freely" [31]. Plessner's conference of 1932 (*Rebirth of Form in the Technical Age*) is precisely the confutation of this assumption. Since the industrial revolution, the technical world has taken on an "intrinsically incomplete and open character" [32] with regard to its products and to space and time, which has resulted in a new feeling of infinity, openness and indeterminacy. This situation must, according to Plessner, find an expression, a new architectural form which, developing out of the needs of technical production, must hold itself in balance between practical and functional needs and playfulness, experimentation and aesthetic considerations: an open form. The new form must renounce symmetry and completeness and tend towards its own transcendence and transformation, rendered possible by the use of new materials such as concrete and glass allowing for a potentially infinite growth of the initial structure.

The same argument is discussed by Umberto Eco in the introduction to the first Italian edition of *Opera Aperta*, that is, a series of essays concerning, as Eco says, the reaction of art and artists to the proving of Chance, of the Undetermined, of the Probable, of the Ambiguous, of the Plurivalent, that is, the reaction of art to the new epistemological horizon that mathematics, biology, physics, psychology and so on have opened up [33]. As Eco says, the open work is more a hypothetical model than a reality; it depicts a tension, a problem. Even if, in general, every work of art is supposed to tend towards "openness", that is, it generates a personal response as the fundamental condition of the aesthetic experience [34], contemporary art is, according to Eco, confronted with the end of the traditional cosmological order that was immutable and objective. This disorder requires the attempt to give it a form, not in the sense of seeking to master or imprisoning ambiguity, but in the sense of elaborating new forms of relationships in which ambiguity finds its own justification and acquires a positive value. The poetic of the "open" work is thus to "encourage 'acts of conscious freedom' on the part of the performer and place him at the focal point of a network of limitless interrelations, among which he chooses to set up his own form without being influenced by an external necessity which definitively prescribes the organisation of the work in hand" [35]. The difference with any other work of art is that in (Eco's) contemporary art the "openness" has been subsumed into a positive and conscious aspect of artistic production, thus becoming an explicit feature.

Regarding the debate generated by Eco's book, we will only cite Bruno Zevi's position on the "open work" and architecture. The poetics of the open work, as Zevi defines it, is indeterminateness, ambiguity, plurivalence, disorder. It is "the rejection of the classical concepts of continuity, universal law, causal relationship, predictability of phenomena, general formulas capable of 'explaining' the world, life or history. [...] The composer no longer creates in a definitive and conclusive manner, guiding the performer or the spectator in an unambiguous direction, but limits himself almost to handing over the pieces [...]" [36]. While in the seventeenth century this poetic was dictated by the programme of persuading and propagandising in the context of authoritarian societies, the open form is, in contemporary times, caught in a dialectic between revolutionary and reactionary implications. Thus, in itself, it offers "no guarantee of being on the revolutionary or at least protesting wave [since] neo-capitalism accepts the 'open' work in any factory or office block" [37]. This is why, for Zevi, it is necessary to verify the validity of its contents and their expressions.

More recently, Richard Sennett has taken up, with some adjustments, Jane Jacobs's critique of Le Corbusier's urban vision, considering the application of the open form to urban planning and architecture as the means for a tangible experience of democracy. In this regard, Sennett imagines a series of material interventions that are based, for example, on the design of incomplete, non-linear and porous spaces [38] (see *Grenze* versus *Schranke*).

References

1. Ritter, J. (eds).: Historisches Wörterbuch der Philosophie, vol. 3, pp. 875–877, Schwabe & Co. Verlag, Basel/Stuttgart (1974)
2. Gentile, A.: "Limiti" e "confini" della ragione, p. 181. Archivio di Filosofia **76**(3), 179–187 (2008)
3. Porcheddu, F.: Grenze und Schranke: l'origine del soggetto kantiano, p. 364. Areté **5**, 359–379 (2020)
4. Priest, G.: Beyond the Limits of Thought, p. 3. Cambridge University Press, Cambridge (1995)
5. *ivi* p. 110, 111
6. *ivi* p. 81
7. Kant, I.: Prolegomena to Any Future Metaphysics That Will Be Able to Come Forward as Science, p. 104. Cambridge University Press, Cambridge (2004, first published in 1783)
8. Deligiorgi, K.: Kant, Hegel, and the Bounds of Thought, pp. 64–65. Hegel Bulletin **23**(1–2), 56–71 (2002)
9. König J.: Der Begriff der Intuition, Georg Olms Verlag, Hildesheim–New York (1981, first published in 1926)
10. Holz, H.H.: Mensch–Natur. Helmuth Plessner und das Konzept einer dialektischen Anthropologie, p. 39, transcript Verlag, Bielefeld (2003)
11. Plessner, H.: Levels of the Organic Life and the Human. An Introduction to Philosophical Anthropology, p. xxxii. Fordham University Press, New York (2019, first published in 1928)
12. *ivi* p. 96
13. *ivi* p. 97, 98
14. *ivi* p. 145
15. Bernstein, J.M.: Introduction, p. xlix. In: Plessner, H.: Levels of the Organic Life and the Human. An Introduction to Philosophical Anthropology, pp. xxxvii–lxv. Fordham University Press, New York (2019)
16. Buytendijk, F.J.J.: Anschauliche Kennzeichen des Organischen. In: Buytendijk, F.J.J.: Das Menschliche. Wege zu seinem Verständnis, pp. 1–13. Koehler, Stuttgart (1958; first published in 1928)
17. Simmel, G.: Lebensanschauung. Vier metaphysische Kapitel, Duncker & Humblot, Berlin (2018) online at: https://socio.ch/sim/lebensanschauung/index.htm
18. Simmel, G.: Sociology: Inquiries into the Construction of Social Forms, p. 551, Brill, Leiden/Boston, (2009, first published in 1908)
19. Fitzi, G.: Il concetto di "limite" come categoria fondativa dell'antropologia filosofica. Un confronto fra Plessner e Simmel, p. 260. In: Borsari, A., Russo, M.: Helmuth Plessner. Corporeità, natura e storia nell'antropologia filosofica, pp. 245–260, Rubbettino, Soveria Mannelli (2005)
20. Plessner, H.: Political Anthropology. Northwestern University Press, Evanston, Illinois (2018, first published in 1931)
21. Sennett, R.: Building and Dwelling, p. 220. Farrar, Straus and Giroux, New York (2018)

22. Wölfflin, H.: Principles of Art History. The Problem of the Development of Style in Early Modern Art, p. 204, Getty Research Institute, Los Angeles (2015, first published in 1915)
23. id
24. *ivi* p. 205
25. *ivi* p. 206
26. *ivi* p. 214
27. Plessner recognises that Driesch does not ascribe absolute significance to this opposition because, for example, he finds "open forms" in the animal kingdom as well. In: Plessner, H. (ed.) Levels of the Organic Life and the Human. An Introduction to Philosophical Anthropology, p. 203. Fordham University Press, New York (2019, first published in 1928)
28. *ivi* p. 219.
29. *ivi* p. 191
30. Plessner, H.: Political Anthropology, pp. 25–26. Northwestern University Press, Evanston, Illinois (2018, first published in 1931)
31. Wölfflin, op. cit. p. 228
32. Plessner, H., Blower, J.: Rebirth of form in the technical age, p. 44. Art Transl. **3**(1), 37–51 (2011, first published in 1932)
33. Eco, U.: Opera Aperta, pp. 2–3. Bompiani, Milano (1997, first published in 1962)
34. A work of art, therefore, is a [...] *closed* form in its uniqueness as a balanced organic whole, while at the same time constituting an *open* product on account for its susceptibility to countless different interpretations which do not impinge on its unalterable specificity. Hence, every reception of a work of art is both an *interpretation* and a *performance* of it, because in every reception the work takes on a fresh perspective for itself. Nonetheless, it is obvious that works like those of Berio and Stockhausen are "open" in a far more tangible sense. In primitive terms we can say that they are quite literally "unfinished" [...]. Eco, U.: The Open Work, pp. 3–4. Harvard University Press, Cambridge, Massachussets (1989)
35. ivi p. 4
36. Translated from Zevi, B.: Il linguaggio moderno dell'architettura. Guida al codice anticlassico, p. 151. Einaudi, Torino (1973)
37. *ivi* p. 153
38. Sennett, R.: Building and Dwelling. Farrar, Straus and Giroux, New York (2018)

Claudia Nigrelli is a Ph.D. candidate in Architecture and Design Cultures at Alma Mater Studiorum University of Bologna (XXXVI Cycle), joint Ph.D. program with Ecole des Hautes Etudes en Sciences Sociales, Paris. Her research project, entitled "Forms and Figurations in Helmuth Plessner: life, technique, architecture", seeks to reconstruct the originality of Plessner's reflection on form architecture.

Deliberation/Climate Justice

Marco Palma

Abstract

Representative democracy faces a crisis, and the capitalist economic system has not been able to guarantee the communitarian interest. Deliberative processes could contribute to enhancing democracy and tackling inequality, also through new approaches more linked to the concept of common goods. However, deliberative tools might also introduce new inequalities between people living in different places, making citizenship a temporary and/or more territorially defined right. Climate change is transforming the scenario in which social relations are developed, widening the rift created by inequality. Citizenship is indissolubly linked to national borders, but the consequences of global warming lead us to criticise the capacity of citizenship to guarantee an inclusive answer to the increasing risks related to the climate crisis.

Keywords

Democracy · Commons · Participation · Deliberation · Climate justice · Climate change · Citizenship · Global warming

M. Palma (✉)
Department of Architecture, University of Bologna, Bologna, Italy
e-mail: m.palma@unibo.it

1 Deliberation

Citizenship is associated with sovereignty. Starting from the second half of the twentieth century, sovereignty in European countries expressed itself as representative democracy, which is still the prevalent expression of sovereignty in Europe.

Nowadays, representative democracy faces a crisis, and an increasing number of citizens consider themselves inadequately represented, therefore not 'sovereign' of the decisions that regard their daily life. Many researchers link this crisis to the decreasing quality of democratic institutions' answers to increasing social demands [1].

Furthermore, the capitalist economic system typically associated with representative democracy is based on the dichotomy between private and public goods; this dichotomy has not been sufficient at guaranteeing the communitarian interest. This is made clear from issues related to vital resources like water. Starting with this evidence, a large body of literature has been developed, [2–4] extending the concept of 'common goods' from primary resources to all tangible and intangible goods that influence our everyday life.

In contemporary society, the way in which a community decides how to manage communitarian goods is a substantial element in defining citizenship. In a representative democracy, delegating an elected representative is the way for an

A. Borsari et al. (eds.), *TEMPORARY: Citizenship, Architecture and City*, The City Project 4,
https://doi.org/10.1007/978-3-031-36667-3_22

individual to participate in public goods management; moreover, elected representatives define the legal framework to define and manage private goods. The double crisis described in the previous paragraphs misdirects the institution of delegation, and this breaks the link between sovereignty—intended as the ability to participate in the processes that define the everyday life of people and communities—and citizenship.

Deliberative processes contribute to defining citizenship. The studies on common goods intersect those on deliberative practices, including the object and not only the subject of deliberation. However, considerations made in the previous paragraphs do not have a univocal dimension: they apply differently in specific local, social and political contexts. Different political choices and local social dynamics, the territorial demographic dimension and their cultural, ethnographic and economic profile characterise deliberative processes in different ways, defining a ladder that goes from manipulation to citizens' control [5].

As a consequence, citizenship may take different forms depending on how innovative the participatory practices are. From this point of view, we can affirm that citizenship may also have temporary characteristics for a person who, while maintaining the right to vote—and therefore the right to be represented—in their place of residence, may have the opportunity to participate in deliberative processes in their places of work or study. This should lead to the paradox of diversification of citizenship for the same individual.

In other words, the way decisions are made that influence the public realm as well as collective and private life has direct consequences on the extension of the citizenship concept, as there are different ways to contribute to sovereignty and participate in the communitarian value definition of tangible and intangible goods. In a global context in which relations go beyond community borders, becoming a kind of good to which people have different access, citizenship—known as a universalistic concept—risks to be declined temporarily and territorially, and for those reasons become uncertain.

In further detail, the forms of decision-making process may guarantee more or less complete citizenship, depending on the inclusivity of the selected decision-making procedures. There are several variables influencing the participatory opportunities of each individual [1], and different ways to empirically concretise deliberative democracy [6]; however, we can suggest that citizenship—which is mainly described as a sum of rights and duties—finds its full expression in the participatory opportunities that allow to take part in 'sovereignty'. From that point of view, deliberative democracy processes may represent a temporary recognition of the right to citizenship that an individual might enjoy in a precise time and space in which they have the possibility and want to participate.

Considering deliberative democracy as an aspect of citizenship means to recognise and understand in depth its scarcity and uncertainty. Not just because a deliberative process has a defined territoriality that is different to the one we commonly attribute to the concept, but also because each deliberative process implicates inclusions and exclusions concerning both the subjects and objects of deliberation.

2 Climate Justice

Several factors able to guarantee a minimum quality of life are associated with citizenship. Among these we can list welfare, social and civic rights, and work. However, citizenship alone has not been able to guarantee the minimum equality among individuals envisaged by the principles, and poverty and low living standards still characterise the everyday life of billions of people around the world, including millions of Europeans.

Climate change is transforming the scenario in which social relations are developed, widening the rift created by inequality. On one hand, climate change has consequences on everyday quality of life, even in the framework of single cities and in the local dimension. For example, heatwaves are becoming more common every summer, while blazes and floods indiscriminately affect populations living in poor or rich territories. This dynamic should put all people 'in the same boat', exposing individuals with different salaries and rights to the same risks.

On the other hand, the richest 10% of the global population (around 630 million people) was responsible for 46% of total CO_2 emissions between 1990 and 2015. In the same period, the poorest 50% of the global population maintained its own impact unvaried, being responsible of the 6% of the increased CO_2 emissions [7]. Furthermore, there has been an increase in the number of so-called climate refugees. The depletion of land has often led to increased inequality, with conflict-displaced migrants looking for asylum elsewhere [8]. In 2020, of an estimated 40.5 million new refugees, 30.7 million have been displaced due to climate and environmental disasters [9].

From this point of view, we can observe that economic activities generating profits for some populations—or for some individuals among those populations—often result in tragic consequences thousands of kilometres away, provoking "an unprecedented human rights crisis" [10] affecting innocent communities exposed to increased risks.

Citizenship is indissolubly linked to national borders; however, the data leads us to criticise the capacity of citizenship to guarantee an inclusive answer to the increasing risks related to the climate crisis. On the one side, the enormous disparity in contributing to climate crisis ingravescence further highlights the already known limitations of citizenship, which divide humanity on a passport basis, establishing a kind of 'atmospheric colonialism': between the 1751 and 2014 the United States, Great Britain and Germany alone produced an overall quantity of greenhouse gas emissions six times higher than average global emissions [11].

On the other side, these phenomena create a new dimension of inequality, with increasing parts of the global population becoming victims of drought, famine and extreme weather events; starvation and malnutrition are expected to increase by 20% by 2050 compared to the current level [12]. Between 2030 and 2050, the climate crisis may cause 250,000 deaths every year, mainly in the poorest countries [13].

A first conclusion has to do with the unsuitability of the link between citizenship and state. As described in the previous paragraphs, the richest countries' citizens increase their own consumption by taking advantage of the poorer countries' resources and dumping on them a significant part of the consequences. People living in the most affected countries are pushed to leave their birthplace looking for better life conditions. However, the richest countries fortify their borders, justifying this measure with the protection of the citizenship concept founded on place of birth or residence, thereby contributing to enhancing the inequalities caused by the climate crisis.

On the other hand, there is an increasing divergence between the haves and the have-nots among citizens of the same state, and in all countries the poorest and marginalised communities are the most affected by the consequences of the climate crisis [14]. Heatwaves and pollution are examples: people with higher salaries can protect themselves using technological solutions which are not accessible for wide segments of the population with low incomes and that often live in suburbs or in the most polluted and dismal territories.

As a consequence, citizenship in the climate crisis era needs to have two characteristics, which it currently does not have: global universality, and the possibility to guarantee mitigation and adaptation actions in each territory, independently from economic resources available to people living there. Otherwise, citizenship becomes a privilege.

References

1. Pellizzoni, L.: Cosa significa partecipare, "Rassegna Italiana di Sociologia, Rivista trimestrale fondata da Camillo Pellizzi" 3/2005, pp. 479–514 (2005)
2. Cacciari, P., La società dei beni comuni. Una rassegna, Ediesse, Rome (2011)
3. Lucarelli, A.: Beni comuni. Dalla teoria all'azione politica, Dissensi, Massarosa (LU) (2011)
4. Mattei, U.: Beni Comuni: un Manifesto. Editori Laterza (2011))
5. Arnstein, S.: A ladder of citizen participation. J. Am. Plann. Assoc. **35**(4), 216–224 (1969)

6. Bobbio L.: La democrazia deliberativa nella pratica, in "Stato e mercato, Rivista quadrimestrale" 1/2005, pp. 67–88, 2005. https://doi.org/10.1425/19634
7. Gore, T., Alestig, M, Ratcliff, A.: Confronting Carbon Inequality. Putting Climate Justice at the Heart of the COVID-19 Recovery, Oxfam Report, 2020. https://oxfamilibrary.openrepository.com/bitstream/handle/10546/621052/mb-confronting-carbon-inequality-210920-en.pdf. Accessed 06 Sept 2021
8. Aime, M., Favole, A., Remotti, F.: Il mondo che avrete. Virus, antropocene, rivoluzione. UTET, Milan (2020)
9. Internal Displacement Monitoring Centre, Internal Displacement in a Changing Climate, Global Report on Internal Displacement 2021, 2021. https://www.internal-displacement.org/sites/default/files/publications/documents/grid2021_idmc.pdf. Accessed 08 Sept 2021
10. Amnesty International, Stop Burning our Rights! What Governments and Corporations must do to protect humanity from the climate crisis, 2021, London. https://www.amnesty.org/en/wp-content/uploads/2021/07/POL3034762021ENGLISH.pdf. Accessed 07 Sept 2021
11. Gore T., Alestig, M., Ratcliff, A.: Confronting Carbon Inequality. Putting climate justice at the heart of the COVID-19 recovery, Oxfam Report, 2020. https://oxfamilibrary.openrepository.com/bitstream/handle/10546/621052/mb-confronting-carbon-inequality-210920-en.pdf. Accessed 06 Sept 2021
12. UN World Food Programme, Cambiamenti climatici e fame. Un mondo a Fame Zero ha bisogno di resilienza al clima, 2019. https://docs.wfp.org/api/documents/WFP-0000105744/download/?_ga=2.136993328.1635603569.1631116405-1424256918.1631116404. Accessed 07 Sept 2021
13. World Health Organisation, Climate change and health, 2018. https://www.who.int/news-room/fact-sheets/detail/climate-change-and-health. Accessed 08 Sept 2021
14. David, R., Luporini, R.: Cos'è la giustizia climatica e perché è importante, ISPI, 2021. https://www.ispionline.it/it/pubblicazione/cose-la-giustizia-climatica-e-perche-e-importante-31232. Accessed 13 Sept 2021

Marco Palma is a Ph.D. candidate in Architecture and Design Cultures at Alma Mater Studiorum University of Bologna (XXXVI Cycle). His research, entlited "Towards Europe 2050: Resilient Cities and European Union. The social impact of the research projects funded by European funds", focuses on the impacts of European funded projects aimed at facing climate change in urban areas.

Border/Rules

Dafni Retzepi

Abstract

Through the examination of six examples, both architectural and artistic, the words "border" and "rules" are explained as generative concepts for the creative project, both stemming from the shared impulse of framing and ordering processes and spaces. The "border" can signify first and foremost a line, a material separation of spaces and a basic delimitation of architectural form, but it can equally be a conceptually invisible physical separation between two antithetical relation of that which is divided, and finally, a "border" can affirm its importance through its own absence. The notion of "rules" on the other hand, can be interpreted as a starting point for an automatic process of creation, as a means of collective liberation by orderly framing the singularity of exceptions, or as a mechanism of resistance against itself.

Keywords

Periphery · Perimeter · Limit · Obligations · Order · Objectivity

D. Retzepi (✉)
Department of Architecture, University of Bologna, Bologna, Italy
e-mail: dafni.retzepi2@unibo.it

1 Border

1.1 Noun: The Importance of the Periphery

In this project, located on the United States-Mexico frontier, the familiar definition of the word *border* as a dividing line separating two distinct entities is lightly distorted, by dilating the usual linearity into a delimited rectangular space. The commonly established binary nature of the line is not directly constrained by the project, but rather it is slightly disoriented by introducing an imposed spatial delay. The *border* is thus defined by one interior and two exteriors, and its line is exploited as a means of creating architectural form. The contained space is an area of shared order, an artificial, controlled paradise of transition contrasting both countries' exteriors, an irregularity in the system, an oasis in the desert. The sensibility of the project relies strongly on its singular, non-serial nature, acquiring the status of a pleasant exception able to offer instant relief rather than a generalisable strategy of action. Consequently, the project does not challenge the established reality but stretches its limits while remaining realistic, pragmatic and programmatically faithful. Architecture safely sticks to its proper vocabulary and formal dictionary, proposing a moment of controlled deformation within the existing definition of *border*.

A. Borsari et al. (eds.), *TEMPORARY: Citizenship, Architecture and City*, The City Project 4,
https://doi.org/10.1007/978-3-031-36667-3_23

1.2 Noun: The Importance of the Limit; Interaction Between Exterior and Interior

While acknowledging the peri-urban reality of its context in the region of Westerlo in Belgium, the project actively confronts itself against the condition of urban sprawl by proposing a physical *border*, a linear boundary aimed at protecting the exceptional and rare, non-urbanised qualities of its parcel. All programmatic constraints are thus condensed in one inhabited, delimiting line, retaining the suburban sprawl like a barrage. Although the definition of this *border* is linear, the centre of importance of the project deviates from the architecture of the boundary, towards the void being framed and protected. The contained interior space is non-programmed, non-designed, unordered. The project delicately balances itself on the ambiguity created between the line of the defined *border* of individual inhabitation, and the collective space of shared liberty, and nourishes itself from this contrast. The *border* pushes architecture beyond its line, challenging the established notions of private ownership and social belonging.

1.3 Noun: The Annihilation of the Border; Architecture Without Content

The project for student housing in the University of Paris-Saclay accomplishes a radical dissolution of *border*, a complete blurring of limits. The rational concrete structure of the *Maison Dom-Ino*, visible on the facade, establishes the potentiality of liberty on every level. There is no hierarchy among the spaces. Enclosure happens effortlessly, like sliding pieces of glass on a flat, smooth surface. The context does not seem to have an influence on architecture, neither does the programme, nor the border, nor the notion of citizenship. There is no specificity, no exception, no ambiguity. There is no contrast, no dialogue, the project could expand infinitely. Architecture is confronted to the absence of content. The *border* as the potential of what could be.

2 Rules

2.1 Noun: A Point of Departure; an Automatic Sequence of Events

Citizenship is both dependent and defined by a set of rights and obligations, laws and stipulated *rules* ensuring the orderly coexistence of the individual into the whole. Prescribed by the authorities, these regulations describe norms of social conduct common to all citizens, aimed at achieving a status of equality. *Rules* are conceived in order to regulate human behaviour and prevent the arbitrariness of actions based on self-interest at the expense of the common. Architecture, as a discipline orchestrating the common, is equally defined by *rules*, directed by norms and regulations to be obeyed by all architects, thus guaranteeing a common ground of objective reality.

The notion of *rules* is directly addressed in the *Chris Marker* building. Due to a particularity in the fire safety *rules*,[1] the 100 m-long facade presents characteristic concrete triangular motifs, themselves defining the shape of the pillars, the repetition of which imply a variation of their scale. The particularity of the facade is demonstrated as a self-evident choice, a chain reaction taking place as the mere consequence of a citizen abiding the law. The *rules* generated an order exterior to the architect's will, preventing arbitrariness of decision. *Rules* are the centre of the discourse, their abstraction and artificiality are admitted as qualities. No cheating is observed, only an elegant manipulation of their corners. Creation is thus neutralised and the architect observes from a distance, relying on the *rules*—is the architect the *rules*?

2.2 Noun: The Importance of Order

The project for a new city plan in South Korea consists of a grid superimposed on the landscape, the layout deforming itself according to the present territorial conditions. The grid was materialised architecturally through a series of cruciform buildings forming the basic architectonic structure of the city, while simultaneously providing "rooms without content" in the in-between spaces, becoming "the space for further urban development."[2] In a context where everything was possible, the self-imposed *rule* of the grid establishes order. Its rigorousness provides a controlled and controlling stable grammar which is non-negotiable. The buildings, represented in the images as abstract white forms, intentionally form the background of what is to come. The rigidity of the *rule* controlling the architects' liberty of action paradoxically provides the conditions for a future liberty of appropriation.

2.3 Noun: Rules Against Rules

In his early explorations as an artist, Ellsworth Kelly developed what Yve-Alain Bois has called "strategies of not–inventing, of not–composing, of not taking decisions."[3] These five self-imposed *rules*—transfer, chance, grid, monochrome and silhouette—acted as guidelines for controlling and orientating the artist's gestures away from the subjective creation of form. The new set of rules was developed as a replacement for the existing norms of composition, a radical questioning of individuality and personal perception, and an attack on the glorification of artistic genius. By means of the production of self-regulated form, Kelly assumes the responsibility for breaking the rules by delicately replacing them, disorientating the discourse and creating calculated newness.

Notes

1. In order to avoid the propagation of fire, two fire compartmented floors have to be separated by adequately resistant material in the length of at least one metre after the end of the facade.
2. OFFICE KGDVS, http://officekgdvs.com/projects/#office-19, description of the project, last accessed 2022/06/03.
3. Bois, Y.: Ellsworth Kelly's Dream of Impersonality. The Institute Letter (2013).

Bibliography

1. Bruther.: El Croquis 197. El Croquis. Madrid (2019)
2. Bruther.: 2G 76. Koening Books Ltd. London (2018)
3. Dogma.: El Croquis 208. El Croquis. Madrid (2021)
4. Gargiani, R.: Oasis pour réfugiés désenchantés et visionnaires éclectiques. MATIÈRES 15 (2019)
5. Office KGDVS.: El Croquis 185. El Croquis, Madrid (2016)
6. Van Gerrewey, C.: How Soon Is Now? Ten Problems and Paradoxes in the Work of Dogma. Log, p. 35 (2015)
7. Van Gerrewey, C.: Order, Disorder. Ten choices and contradictions in the work of OFFICE. OFFICE—Kersten Geers, David Van Severen Volume 2 (2017)
8. Bois, Y.: On Ellsworth Kelly. In: The Menil Collection, Conference, Houston (2015)
9. Bois, Y.: Ellsworth Kelly's Dream of Impersonality. The Institute Letter (2013)
10. Declerck, J., Vande Velde, D.: From Realism to Reality. OASE, p. 67 (2015)
11. Lapierre, E., Geers, K., Van Severen, D.: OFFICE versus Éric Lapierre Experience. In: Cité de l'architecture et du patrimoine, Conference (2019)
12. Lapierre, E.: Surrationalism. In: EPFL, Conference (2020)
13. Lapierre, E.: Marvellous architecture: reason as architectural LSD. In: 100 Day Studio: Architecture Foundation, Conference (2020)
14. Office KGDVS. El Croquis 185. El Croquis, Madrid (2016)
15. Van Gerrewey, C.: Cum fundamento in re. Architecture and Reality. OASE, p. 90 (2015)
16. Van Gerrewey, C.: Order, Disorder. Ten choices and contradictions in the work of OFFICE. OFFICE Kersten Geers David Van Severen Volume 2 (2017)

Dafni Retzepi is a Ph.D. candidate in Architecture and Design Cultures at Alma Mater Studiorum University of Bologna (XXXVI Cycle). Her research in the laboratory of architectural theory bears the title "The Corners of Form: Politics at the Boundary. Western Architectural Discourses 1988–2008".

Appendix

Alphabetical Index of Lemmas

Adaptation (*Marco Iannantuono*).
Assemblages (*Andrea Cattabriga*).
Border (*Dafni Kalliopi Retzepi*).
Classroom (*Angelo Massafra*).
Climate justice (*Marco Palma*).
Cueva and Tomas de terreno (*Marco Iannantuono*).
Decentralized (*Andrea Cattabriga*).
Deliberation (*Marco Palma*).
Deviance (*Valentina De Matteo*).
Digital Citizenship (*Angelo Massafra*).
Grenze vs. Schranke (*Claudia Nigrelli*).
Hacktivism (*Francesco Di Maio*).
Life cycle phase (*Lorna Dragonetti*).
Life cycle thinking (*Lorna Dragonetti*).
Material Culture (*Angelo Massafra*).
Open form (*Claudia Nigrelli*).
Pop-up spaces (*Giulia Marzani*).
Rules (*Dafni Kalliopi Retzepi*).
Sexual tourism (*Francesco Di Maio*).
Temporary uses (*Giulia Marzani*).
Time-based packaged goods (*Clara Giardina*).
Transition (*Valentina De Matteo*).
Valigia (*Clara Giardina*).

A. Borsari et al. (eds.), *TEMPORARY: Citizenship, Architecture and City*, The City Project 4,
https://doi.org/10.1007/978-3-031-36667-3

Author Index

A
Ascari, Pierpaolo, 1, 65

B
Borsari, Andrea, 1, 109

C
Cattabriga, Andrea, 123

D
De Matteo, Valentina, 127
Di Maio, Francesco, 131
Dragonetti, Lorna, 137

E
Eghbali, Arshia, 91

F
Fitzi, Gregor, 13

G
Galimberti, Jacopo, 57
Gasperoni, Lidia, 27
Giardina, Clara, 141

I
Iannantuono, Marco, 145

M
Marzani, Giulia, 149
Massafra, Angelo, 153
Massari, Martina, 71
Mubi Brighenti, Andrea, 49

N
Nigrelli, Claudia, 159

P
Pagliula, Serena, 81
Palma, Marco, 167

R
Rahola, Federico, 39
Retzepi, Dafni, 171

T
Trentin, Annalisa, 1, 101

A. Borsari et al. (eds.), *TEMPORARY: Citizenship, Architecture and City*, The City Project 4,
https://doi.org/10.1007/978-3-031-36667-3